USS MISSOURI
(BB-63)

"The Mighty Mo"

TURNER PUBLISHING COMPANY

TABLE OF CONTENTS

Association President's Message 4

Publisher's Message 5

History 6

Special Stories 20

Veterans 42

Biographies 44

Roster 83

Index 94

TURNER PUBLISHING COMPANY

Copyright © 1998
Publishing Rights: Turner Publishing Company
This book or any part thereof may not be
reproduced without written consent of the
publisher.

Turner Publishing Company Staff:
Project Coordinator: John Mark Jackson
Designer: Shelley R. Davidson

Library of Congress Catalog Card No.
98-86843
ISBN: 978-1-68162-452-5

Printed in the United States of America. Additional copies may be purchased directly from the
publisher. Limited Edition.

ASSOCIATION PRESIDENT'S MESSAGE

Dear Shipmates,

At the time of this history book's printing, the USS Missouri has made History once again. On June 22, 1998, our ship found her final resting place in the waters of Pearl Harbor. She will be there for eternity, and forever remind the world of the Alpha and Omega of the World War in the Pacific. The beginning, with the bombing of Pearl Harbor by the Japanese and memorialized by the USS Arizona. The end, with the signing of the Japanese surrender aboard the Mighty Mo and her final resting place in Pearl Harbor near the Arizona.

The stories, photos, and comments in the enclosed pages are but a few of the many that abound. We are sorry that many of our shipmates could not contribute, having gone on tho the kingdom of the deep. We have included all the names of known crew of Missouri, alive and departed. Those of us who carry on the spirit and camaraderie of Missouri, will endeavor to make sure future generation and historians are aware of our ship's contribution to world peace.

Among our membership, we have men who served in the Navy on December 7, 1941, in Pearl Harbor, men who served in the Navy on D-Day on the beaches of Normandy on June 6, 1944, men who were at the commissioning of the ship in the Brooklyn Navy Yard on June 11, 1944. We have men who saw the ship commissioned and decommissioned twice. There are those who were aboard on September 2, 1945, to witness the signing of the surrender of Japan on her 01 deck, bringing to a close World War II. And then we have men who were aboard when the Mighty Mo again went in harms way to Korea in two cruises, and to the Persian Gulf, where she fought her last battle.

The U.S.S. Missouri (BB-63) is the last battleship. The men of Missouri are the last of a breed of sailor, that served our country since the war of the Revolution, the end of an era, that of the Battleship Sailor.

Thank you for the honor of being your President and being a part of this moment in history. Here's to fair winds, following seas and keeping a straight wake.

Herbert "Herb" Fahr, Jr.
President 1997-1998
U.S.S. Missouri (BB-63) Association, Inc.

USS MISSOURI HISTORY

USS Missouri (BB-63) History

The following are significant events in USS Missouri's history.

6 January 1941: Keel laid at New York Naval Shipyard, Brooklyn, NY.

29 January 1944: Launched under sponsorship of Miss Margaret Truman

11 June 1944: Commissioned at New York Naval Shipyard, Brooklyn, NY.

14 December 1944: Departed San Francisco for entry into World War II War Zone in the Pacific.

15 February 1945: Participated in raids against Iwo Jima.

10 March 1945: Participated in raids against Okinawa.

29 August 1945: Arrived Tokyo Bay.

2 September 1945: Instrument of Formal Surrender signed aboard USS Missouri, Tokyo Bay, Japan.

27 October 1945: President Harry S. Truman visits USS Missouri in New York Harbor on Navy Day.

5 April 1946: Port visit to Istanbul, Turkey, to deliver remains of Turkish Ambassador who died in Washington, DC during the war.

30 August 1947: Presidential cruise to Rio de Janeiro; President Truman, Mrs. Truman and Miss Margaret Truman embarked.

19 August 1950: Departed Norfolk, Virginia en route to Korean Theater.

11 September 1952: Departed for second deployment to Korean Theater.

26 February 1955: Decommissioned; entered reserve fleet, Bremerton, Washington.

14 May 1984: Under tow from Bremerton, Washington, en route to Long Beach Naval Shipyard.

10 May 1986: Decommissioned by Secretary of Defense Casper Weinberger, San Francisco, California.

During 1986 through 1989, USS Missouri participated in an around the world cruise, the first battleship to circumnavigate the world since President Theodore Roosevelt's "Great White Fleet." Missouri journeyed to the Persian Gulf near the Strait of Hormiz for supporting operations, participated in the Rim of the Pacific exercises and visited Pusan, Korea.

In September 1990, USS Missouri deployed to the Persian Gulf in support of Desert Shield. On the first day of Desert Storm, she fired her 16" guns for the first time in anger for over 35 years.

In November 1993, USS Missouri departed Long Beach for Pearl Harbor, where she was the host ship for the 50th anniversary of the attack on Pearl Harbor. She returned to Long Beach on her final cruise to be decommissioned on 31 March 1992.

The USS Missouri (BB-63) now rests in the Bremerton Naval Shipyard as part of the Reserve Fleet. Upon call, Missouri will still be a powerful and fearful dreadnought in the best tradition of the US Navy.

5 January 1995 - The USS Missouri and her sister ships, Wisconsin, Iowa and New Jersey, are stricken from the Naval Register by John Dalton, Secretary of the Navy.

2 September 1995: History is repeated as the "Mighty Mo" is honored by the USS Missouri Association (BB-63), Inc. with over 600 members of the original crew holding a Memorial Service on her fantail. The original ship's Band led by Chief Tom Hill, play the themes for the service. Guest speakers include Captain Joe Taussig who guided the USS Nevada out of harms way at the Pearl Harbor attack, General Richard Hearney, USMC, Horace Dicks, and Barbara London, woman pilot of WWII. Also speaking were the association officers, Vice President Herbert Fahr Jr. and Reunion Chairman Jack Hynes. Almost 8,000 visitors witnessed the historic dockside event in Bremerton, Washington.

21 August 1996: Battleship Missouri to be donated to Memorial Association in Hawaii. After a lengthy debate by San Francisco, Bremerton, Long Beach and Honolulu, the Mighty Mo will be going to Hawaii where she will become a memorial alongside the USS Arizona memorial, and become "where WWII started with the Pearl Harbor attack and where WWII ended, on the decks of the USS Missouri."

The ship departed Bremerton, WA under tow, for her last sea journey on May 23, 1998. She stopped in Astoria, Oregon in the Columbia River to have her bottom washed by the fresh water of the river. On June 3, the Crowley "Sea Victory" tugboat with a crew of seven, pulled the ship away and towed Missouri 2,368 miles to Honolulu, arriving on June 21 and docking on June 22. Upon arrival, the ship was handled by volunteers who immediately started a clean-up and painting project to prepare the ship for

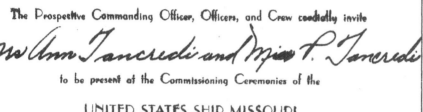

The Prospective Commanding Officer, Officers, and Crew cordially invite

Mrs Ann Tancredi and Miss P. Tancredi

to be present at the Commissioning Ceremonies of the

UNITED STATES SHIP MISSOURI

Sunday the Eleventh of June
Nineteen Hundred and Forty Four
at three o'clock

Navy Yard, Brooklyn, New York

Guest of *D.F. Tancredi*

Ceremony at the debarkation of the body of the Turkish ambassador.

Postage stamp issued by Turkey to commerate the returning of the Turkish Ambassador's body by the USS Missouri.

U.S.S. Missouri

an exceptional 53rd surrender anniversary ceremony to be held on September 2. The ship is temporarily berthed on pier Foxtrot 5 on Ford Island. The USS Missouri (BB-63) Association, Inc. will hold their 1998 reunion in Hawaii with the ABA. The September 2nd ceremony will include many original crew members, plank owners, and witnesses to the surrender signing. The USS Missouri Memorial Association will open the ship to the general public as a Memorial and Museum in January, 1999.

The following are the battle ribbons won by the USS *Missouri*:

• Combat Action Ribbon - For active participation in ground or surface combat subsequent to 1 March 1961, while captain, colonel or junior.

• Navy Unit Commendation - For outstanding heroism in action or extreme meritorious service not involving combat but in support.

• Meritorious Unit Commendation - For valorous and meritorious achievement or service that renders the unit outstanding compared to other units.

• Navy "E" Ribbon - To denote permanent duty on ships that won battle efficiency subsequent to 1 July 1974.

• China Service Medal - For dates served in China, 7 July 1937 to 7 September 1939 and 2 September 1945 to 1 April 1957.

• American Campaign Medal - WWII.

• Asiatic-Pacific Campaign Medal - WWII.

• Victory Medal - WWII.

• Navy Occupation Service Medal - During and after WWII to 4 June 1954.

• National Defense Service Medal - For service 27 June 1950 to 27 July 1954; 1 January 1961 to 14 August 1974, to present.

• Korean Service Medal - For service 27 June 1950 to 27 July 1954.

• Armed Forces Expeditionary Medal - For designated operations after 1 January 1958.

• Southwest Asia Medal - Gulf dates of 2 August 1990 to end.

• Sea Service Deployment Ribbon - To recognize the unique and demanding nature of sea service and the arduous duty attendant with deployment subsequent to 15 August 1974.

• Republic of Korea Presidential Unit Citation - Service in Korea from 27 June 1950 to 27 July 1953.

• United Nations Service Medal - Service in Korean area in support of the UN action from 27 June 1950 to 27 July 1954.

• Liberation of Kuwait Medal - Gulf action dates from 3 August 1990 to end.

Compiled and edited by Herbert Fahr Jr., President of the USS Missouri (BB-63) Association, Inc. 1997-1998.

The USS Missouri (BB-63) Association, Inc. was formed in 1974 and is made up of all former sailors and marines who served aboard her as ship's company. Also invited are all flag members as well as visiting dignitaries. The Association holds an annual reunion on the Labor Day weekend every year to commemorate the 2 September 1945 date. There are also four planning meetings during the year in various locations. Former shipmates are invited to join. For an application, call or write or e-mail. Herb Fahr, 24 Clark Street, Plainview, NY 11803-5114, 1 516 931-1769, MoBB63Mo@aol.com.

Halsey shoe receipt.

Pope Pius XII, Father O'Connor, and Captain Roscoe Hillenboetter.

USS Missouri, launched January 29, 1944 at the New York Navy Yard.

Miss Mary Margaret Truman, sponsor of the USS Missouri, with her father.

History Of The USS *Missouri*
(BB-63)

The USS *Missouri* was built by the Navy Yard, New York, her keel being laid 6 January 1941. She was christened by Miss Margaret Truman on 29 January 1944, and placed in full commission by the Commandant Navy Yard, New York on Sunday, 11 June 1944. Captain William M. Callaghan, USN, accepted the ship and assumed and command.

The ship remained in New York Harbor until 3 August 1944, then operated in Chesapeake Bay until 21 August 1944. On that date the USS *Missouri* departed for the Gulf of Paria, Naval Operations Base, Trinidad, BWI, arriving on 25 August 1944. The ship conducted gunnery, flight, engineering and other shakedown exercises in the area until 17 September 1944. *Missouri* then returned to New York. The ship remained in New York Harbor until final departure with Task Group 27.7 on 11 November 1944 for Cristobal Canal Zone. Transited the Panama Canal and arrived in Balboa on 18 November on which date the ship joined the Pacific Fleet. Departure from the Panama Canal Zone was in company with Task Unit 12.7.1 on 19 November and the ship arrived in San Francisco Bay on 28 November. The USS *Missouri* escorted by the destroyers *Bailey* and *Terry* departed San Francisco on 18 December 1944 as Task Unit 12.7.1 and entered Pearl Harbor, T.H. on 24 December 1944. The USS *Missouri* as part of Task Unit 12.5.9 departed Pearl Harbor on 1 January 1945 headed westward. On 13 January 1945 the *Missouri* arrived at Ulithi, Western Caroline Islands and reported to Commander Third Fleet for duty and on 26 January to Commander Fifth Fleet. The ship operated from Ulithi conducting provisioning and training exercises until 10 February 1945.

10 February to 5 March 1945

The ship departed Ulithi Anchorage on 10 February 1945 in Task Group 58.2 and operated in Task Force 58 during the period from 10 February to 5 March in preparation for and support of the Iwo Jima operation. As part of Task Force 58 the ship participated in the first East Carrier Task Force strikes against Tokyo on 16 and 17 February 1945. The anticipated opposition to these strikes did not materialize. However, on the evening of 19 February, while steaming off Iwo Jima, several small groups of unidentified aircraft were discovered by radar to be closing the formation. The ship opened fire on one of these targets and an enemy aircraft tentatively identified as a "Helen" burst into flames and crashed for a successful conclusion to the ship's first action against the enemy.

The ship participated as part of Task Force 58 in the 19 to 23 February air strikes in support of the landing forces on Iwo Jima, the 25 February strikes against the Tokyo area and the 1 March 1945 strikes against Okinawa Shima.

5 to 13 March 1945

As part of Task Force 58, the ship remained at anchor in Ulithi Anchorage engaging in routine repairs and replenishment from 5 to 13 March. On 9 March the ship was reassigned from Task Group 58.2 to Task Group 58.4.

The ship departed Ulithi Anchorage on 14 March as part of Task Force 59 and following exercises in company with Battleship Squadron Two on 14 and 15 March, the *Missouri* joined Task Group 58.4 on 16 March. As part of Task Force 58 the ship participated in the 18 and 19 March carrier aircraft attacks against Kyushu and the Island Sea area. During the afternoon and night of 17 March enemy aircraft were known to be in the vicinity of the Task Force, however, none closed to within range of the ships of the formation. At 0741 on 18 March an enemy plane succeeded in dropping a bomb on the USS *Enterprise* which was in formation off the *Missouri's* port bow. At 0805 this ship together with others in the formation opened fire at an enemy plane identified as a "Nick" or "Helen." The plane burst into flames and unsuccessfully attempted to crash the USS *Intrepid*. At 0828 and 0850 the ship opened fire on enemy planes. The first was observed to be damaged when the ship ceased fire and was later splashed by the Combat Air Patrol while the second was downed by gunfire. At 1316 the *Missouri* opened fire at a plane which dropped a bomb near the USS *Yorktown* and at 1320 fired upon a plane which approached to 2,500 yards. Both of these planes were destroyed by gunfire. A number of enemy planes remained out of range in the vicinity of the formation until 2115 when the last plane of the day was splashed by a night fighter.

On 19 March eight enemy raids were tracked by radar before sunrise but none closed to within range. At 0708 firing was seen on the horizon and almost immediately a carrier in Task Group 58.2 was seen to burst into flame. This carrier was later identified by TBS as the USS *Franklin*. During the balance of the day there were a number of alerts and enemy planes were downed by the Combat Air Patrol but none approached within range of the formation. During the period 19 to 21 March there were numerous reports of enemy aircraft in the area, however, these were either accounted for by the Combat Air Patrol or did not approach within range of *Missouri's* guns.

On 24 March the ship, with others, was detached from Task Group 58.4 to form Task Force 59. As part of Task Force 59 the ship participated in the bombardment of southeastern Okinawa Shima on 24 March. This was accomplished at extreme range and accurate assessment of damage was therefore, not possible. Thereafter the ship fueled and rejoined Task Group 58.4 on 26 March 1945 and as part of Task Force 58 the ship continued to operate off Okinawa Gunto and participated in strikes against Kyushu until 6 May. During this period there were frequent alerts and enemy aircraft were destroyed by Combat Air Patrol in the vicinity. The ship opened fire on 29 March 1945 on a plane which unsuccessfully attempted to dive upon the USS *Yorktown* and on 7 April the ship was with Task Force 58 during the air strikes which sank the Japanese battleship *Yamato*.

On 11 April 1945, Task Force 58 was engaged in neutralizing sweeps against southern Kyushu airfields. During the morning one enemy raid was destroyed by the Combat Air Patrol. At 1330 several groups of unidentified planes were reported approaching the formation.

USS Missouri at Pusan, Korea; manning the rails for President Syngman Rhee.

4th Division, May 1946.

By 1340 reports had been received that 13 enemy planes had been splashed and that three others were approaching the formation at high speed and low altitude. At 1442 the ship opened fire on a low flying "Zeke" and although many hits were observed, the pilot succeeded in crashing the side of the *Missouri* immediately below the main deck at frame 169 on the starboard side. Parts of the plane were scattered along the starboard side of the ship and the pilots mutilated body landed aboard. One wing of the plane was thrown forward and lodged near 5-inch mount number 3 where gasoline started a fire which was rapidly extinguished. The ship sustained only superficial damage and none of the ship's company was injured. Later during the day the ship unsuccessfully fired upon a twin engine plane which passed approximately 12,000 yards astern of the ship. Enemy planes were known to be in the vicinity during the night and at 2327 the ship commenced firing at a twin engine plane which crashed approximately one minute later. On the next day, ships on the other side of the formation fired upon one enemy plane and enemy snoopers were in the vicinity during the period from 12 to 14 April 1945, but the *Missouri* did not open fire.

On 16 April Task Force 58 was again conducting raids in support of the landing forces on Okinawa Shima and strikes against the Japanese airfields on southern Kyushu. At 0038 the first Japanese planes approached the formation but retired after being fired upon by ships of the screen. From this time until 1303 numerous reports of enemy planes were received but none closed to within range. At 1303 a group of planes which later developed to be kamikazes were discovered heading for the formation. Shortly after 1326 the ship opened fire on a low flying "Zeke" which crashed close aboard the USS *Intrepid*. Two minutes later fire was opened on a second "Zeke" and when hit the pilot of this plane attempted to crash the *Missouri*. The wing tip of this plane struck the ship's aircraft crane on the stern and the "Zeke" crashed a short distance astern exploding violently. Debris was thrown aboard ship but only minor material damage was sustained. At 1335, nine minutes after the ship opened fire on the first plane, a third plane identified as a "Hamp" was fired upon while diving on the ship. The "Hamp" burst into flame, passed over the ship at an altitude of about 300 feet and crashed close aboard off the starboard bow. One minute later two planes dove on the USS *Intrepid*. One succeeded in crashing her and the other was destroyed. From 1514 to 1516 the ship fired upon two planes. One of these crashed for-

MISSOURI TWICE HAD CLOSE CALLS

Suicide Flyers Struck Near Battleship Off Okinawa

Washington, Aug. 28 (AP)—The Navy disclosed today that the battleship Missouri, scene of the forthcoming surrender ceremony in Tokyo Bay, had at least two narrow escapes from Japanese suicide planes.

On April 11, as the 45,000-ton dreadnaught was helping support the Okinawa campaign, she was attacked by a low-flying *kamikaze*. Gunfire from the ship threw the plane off balance and it crashed against the starboard side aft. Gasoline and debris were thrown over the main deck. Fire broke out but was quickly extinguished.

Did Not Alter Course

The big ship never so much as altered her course, and continued on her assigned duty.

Five days later she was again subjected to attempted suicide crashes but none was successful.

Since her initial action, when she escorted carriers of Admiral Marc Mitscher's Task Force 58 in the first mass air strikes against Japan on February 16 and 17, the Missouri followed the tide of battle to the final Japanese surrender.

Support At Iwo Jima

She furnished close support to carriers operating against Iwo Jima in the early days of that campaign, and it was on the evening of February 19 off Iwo Jima, that the Missouri first opened fire on the enemy.

Her radar picked up an approaching enemy plane, and long before the plane was in range, the ship's 5-inch guns were on the target. When the plane came into range, the guns opened up and the crew saw their first enemy hit the water. Following the Iwo Jima operation, the Missouri returned with carriers for more strikes against Tokyo.

Bombarded Okinawa

On March 24, the Missouri, along with other battleships, opened up the first bombardment of Okinawa. Her 16-inch guns knocked out eight coastal defense installations, destroyed several control towers, exploded an ammunition dump and destroyed and damaged several large buildings.

The $100,000,000 battlewagon was built by the New York Navy Yard and commissioned June 11, 1944, nine months ahead of schedule. President Truman, then Senator from Missouri, made the principal address at the ship's launching.

He said then "the time is surely coming when the people of Missouri can thrill with pride as the Missouri and her sisters sail into Tokyo Bay."

Ship's Service Division.

ward of the *Intrepid* and the other close aboard a destroyer. Two minutes later a third plane which passed 6,000 yards astern of the ship was fired upon and disappeared over the horizon. Shortly thereafter a plane was observed to crash and burn in that general direction. During the remainder of the afternoon planes were shot down by other Task Groups but none came within range of the ship. At 2050 and 2110 the ship opened fire on planes which came within 5-inch gun range and both immediately withdrew. Enemy planes dropped window in the vicinity during the balance of the night but none closed the formation.

On 17 April a 35 plane raid was destroyed by the Combat Air Patrol approximately 60 miles from the formation. However, no enemy planes closed the formation. During the night the ship had a surface radar contact which was later developed by destroyers of the screen and resulted in a kill on an enemy submarine on the following day. There was no enemy activity from April 23 to April 28.

On 29 April enemy aircraft was reported destroyed by the Combat Air Patrol in the morning. At 1645 the ships of the formation including the *Missouri* fired upon and downed one enemy plane. Later during the early morning of 30 April, night fighters splashed enemy planes in the vicinity of the formation but no ships fired during the day.

On 1,2 and 3 May, no enemy planes were known to be in the area and on 4-5 May, although Japanese planes were splashed by the Combat Air Patrol, none approached the formation. On 6 May the *Missouri* was detached from Task Group 58.4 and proceeded to Ulithi Anchorage Fleet. The ship arrived in Ulithi on 9 May and remained there until 17 May. On 14 May Captain W.M. Callaghan, USN, was detached from duty as commanding officer of the *Missouri* and was relieved by Captain S.S. Murray, USN. The ship departed Ulithi on 17 May and arrived Apra Harbor,

Guam on 18 May where, at 1527 Admiral W.F. Halsey, USN, Commander Third Fleet, hoisted his flag aboard the USS *Missouri*.

The ship and screening destroyers *McNair* and *Wedderburn* formed Task Group 30.1 on 21 May and departed Apra Harbor for Hagushi Anchorage, Okinawa Shima, arriving 26 May. While at Hagushi Anchorage on 26 May the ship was twice alerted for air attacks but none developed in the immediate vicinity. The ship departed Hagushi Anchorage in the afternoon of 27 May and conducted a bombardment of targets on southeastern Okinawa Shima in support of the occupying forces, and then proceeded to rendezvous with Task Force 38 off eastern Okinawa Shima. At midnight of 27 May, command of all forces of the Fifth Fleet passed to Commander Third Fleet. The *Missouri* rejoined Task Group 38.4 on 28 May. The Task Force remained off Okinawa Gunto with the carriers furnishing air support to the occupation forces. There was no enemy air activity in the vicinity of the Task Force from 28 May to 10 June although during this period the force again conducted strikes on 2 and 3 June against the Kyushu airfields. On 4 June reports of a typhoon 50 miles south southwest of the Task Force were received and the Task Force withdrew from position in the path of the typhoon. Heavy weather was experienced during 5 May and very minor damage was sustained by the ship due to the heavy seas. On 8 June the Force returned to strike southern Kyushu airfields and on 9 and 10 June air strikes were made against the islands of Daito Shoto. On 10 June Task Force 38 commenced retiring to San Pedro Bay, Leyte, Philippine Islands, arriving on 13 June 1945. The period 14 June to 1 July was spent in upkeep, provisioning and recreation at Leyte Anchorage.

1 July to 15 August 1945

The *Missouri* departed Leyte on the morning of 1 July and the first eight days at sea were spent in exercise periods under Unit, Group and Task Force Commanders, while the Task Force was heading in a general northerly direction. On the evening of 9 July a high speed run toward the Tokyo area commenced. At 0400 on the 10th the various air strikes against airfields in the Tokyo area commenced and although enemy aircraft were reported, none succeeded in getting through the air patrol. The Task Force proceeded northeast on 11 July and on 13 July was off northern Honshu and Hokkaido prepared for air strikes which it developed could not be made on account of poor weather and low visibility. On the 14th the air strikes against northern Honshu and Hokkaido shipping and airfields were made. On 15 July the *Missouri* joined Task Unit 34.8.2 for the bombardment of industrial targets located in Muroran Hokkaido. No opposition developed during the approach, nor was there return fire from shore while the Task Unit shelled the Nihon Steel Works and Wanished Iron Works between 0935 and 1027 (Item) with good results. The *Missouri* rejoined Task Group 38.4 in the evening and proceeded south to fuel on 16 July. The Task Force was in position on 17 July to conduct air strikes against airfields in the Tokyo area. However, the weather was again unfavorable for air operations. In the afternoon of the 17th the *Missouri* again joined Task Unit 34.8.2 and proceeded to bombard the Hitachi area, Honshu. There was again no opposition to the approach of the bombardment group and no return fire during the bombardment of industrial targets in the Hitachi area from 2315 on 17 July to 0600 on 18 July. The bombardment was conducted in exceedingly poor weather which made spotting or illumination of targets as well a determination of the bombardment results impossible.

On 18 July the *Missouri* rejoined Task Group 38.4 which conducted air strikes against targets in the Tokyo area on that day. During 20, 21 and 22 July the most extensive replenishment of fuel, ammunition and provisions were attempted at sea was completed and on 23 July the Task Force again was en route for strikes against combatant shipping in the Kure-Kobe area of the Inland Sea and although enemy planes were reported in the vicinity none succeeded in evading the Combat Air Patrol. Poor weather had prevailed during these strikes and they were therefore repeated on 28 July, again with no enemy air activity over the Task Force. On the 29th a return to the Tokyo area commenced and on 30 July aircraft of the Task Force hit the Tokyo-Nagoya area. Again there was no enemy air opposition over the Task Force. The first six days of August were spent in fueling and maneuvers to avoid the paths of two typhoons which moved north along the Japanese coast. On 7 August the Task Force commenced a run to position to strike northern Honshu and Hokkaido, however, on 8 August fog and low visibility prevented flight operations and the Task Force proceeded south in search of more favorable weather. On 8 August Japanese aircraft were encountered by the Combat Air Patrol and on 9 August the picket destroyers of the formation had been under attack and at 1610 a "Grace" was splashed astern

AV Division, USS Missouri, December 23, 1945. Front: Mr. Whittiker, Robert Hall, Douglas Anderson, H.M. Murphey. Middle: Hugh Kirby Riner, Jack Revels, Ralph W. Conners, Noel Crockett, William Pieland, Leonard Otis Toombs, W.C. Haberstroh, John E. Walsh, Everett N. Frothingham. Back: Dwight H. Williams, Jack Dillon, Jake Jacobson, Bob Muntz, Steve Gabney, E.J. Morrisey, Ed Sadowsky, and Orville Foster.

Commander Third Fleet Flag Marines, USS Missouri, August 1945.

13

Nimitz signs surrender document.

General MacArthur aboard the USS Missouri explaining that the surrender terms will be strictly enforced.

of the *Missouri* and close aboard the USS *Wasp*. Due to the *Missouri's* position in the formation the 40mm guns only were able to fire at this plane. The 10th to 12th of August were spent in replenishment and many conferences of Task Force and Group Commanders were held aboard the *Missouri* as a result of the information received concerning Japanese surrender proposals.

On 13 August other Task Groups of Task Force 38 were under air attack but no enemy aircraft were over Task Group 38.4. 14 August was spent in getting into position for further strikes against the Tokyo area. These strikes were launched on 15 August but were recalled as a result of an urgent dispatch from CincPac. At 1109, by direction of Commander Third Fleet the *Missouri's* whistle and siren were sounded for a period of one minute while battle colors were broken and Admiral Halsey's personal flag was raised in official recognition of the end of hostilities against the Japanese Empire. During this day the Combat Air Patrol splashed Japanese aircraft in the vicinity of the Task Force but none penetrated the patrol.

From 15 to 26 August the *Missouri* operated off the coast of Japan awaiting orders to proceed with the occupation of Japan. On 27 August the Missouri and escorting destroyers proceeded into Sagami Wan, Honshu, having taken aboard Japanese emissaries and a pilot. The 28th of August was spent at anchor and on 29 August the *Missouri* got underway and entered Tokyo Bay anchoring off Yokosuka Naval Station at 0925.

The ship remained at anchor in Tokyo Bay without incident, until 2 September on which day the formal document of the Japanese surrender was executed aboard the USS *Missouri*. On that day Fleet Admiral C.W. Nimitz boarded the *Missouri* at 0805 and his personal flag was broken. At 0843 General of the Army Douglas MacArthur came aboard. At 0856 the Japanese representatives arrived and between 0902 and 0906 the Japanese representatives signed the Instrument of Surrender and two minutes later General MacArthur signed the Instrument. The ceremony was completed at 0925 and the various dignitaries departed the ship. Thereafter the *Missouri* remained at anchor in Tokyo Bay until 6 September 1945, when she departed for Apra Harbor, Guam. Admiral William F. Halsey trans-

ferred his flag as Commander Third Fleet to the USS *South Dakota* on 5 September 1945. Passage from Tokyo Bay to Guam was without incident and the *Missouri* arrived in Apra Harbor on 9 September. The ship departed Guam with homeward bound veterans on 12 September 1945 and arrived Pearl Harbor, T.H. on 20 September 1945.

Post World War II

On 29 September 1945, *Missouri* departed Pearl Harbor and headed for the eastern seaboard of the United States. Transiting the Panama Canal, she headed for New York where she became the flagship of Admiral Jonas Ingram, Commander in Chief, United States Atlantic Fleet, on 24 October 1945. On 27 October 1945, the *Missouri* boomed out a 21 gun salute as she was boarded by President Harry S. Truman during Navy Day celebration ceremonies.

After overhaul in the New York Navy Yard, and a training cruise to Cuba, the *Missouri* was on her way to Gibralter in March 1946. From there she passed into the Mediterranean on a goodwill mission that served also as an impres-

sive demonstration of American military power. Her presence symbolized United States support for the rights and freedom of Greece and Turkey, both in danger on being drawn into the Soviet orbit of satellite states.

In Rio de Janeiro, on 2 September 1947, the *Missouri* was again a symbol of American strength in support of its Allies against the advances of Communist aggression. The *Missouri* provided the site for President Truman to sign the Rio Treaty which made the Monroe Doctrine a multilateral pact. Business and ceremonial duties concluded, President Truman, accompanied by Mrs. Truman and his daughter Margaret, returned to the United States aboard the battleship. From 23 September 1947 to 10 March 1948, the *Missouri* was in the New York Navy Yard for overhaul and then went on a training cruise to Guantanamo Bay, Cuba. She arrived in Annapolis in June to take on midshipmen for a training cruise to Portugal, France, Algeria and back to Cuba.

On 17 January 1950, heading to sea from Hampton Roads, the *Missouri* ran aground. It was 0825, close to high tide, when the battleship ran aground 1.6 miles from Thimble Shoals Lights

GREETINGS
From the Flagship
USS Missouri in Tokyo Bay

We have reached the zenith of our triumphs and the Stars and Stripes fly over Tokyo Bay today. Since December 7, 1941, we have had a long, grim, bloody struggle, and untold sacrifices. Great acts of courage and valor have been shown by the men of the U.S. Armed Forces in reaching our goal---the Official Surrender of the Japanese Forces, on September 2, 1945.

Post card sent home by crew members.

near Old Point Comfort. She traversed shoal water a distance of three ship lengths, about 2,500 feet, from the main channel. Lifted about seven feet above the water line, she stuck hard and fast. It took many tugs, pontoons, and an incoming tide to free her finally on 1 February. The incident provided Navy personnel with valuable experience in extensive and diverse salvage work.

Korea

Until called to support United Nations Forces in embattled Korea in 1950, the *Missouri* trained thousands of naval reserves, midshipmen, and other naval personnel on cruises from New England to the Caribbean and across the Atlantic to English and European waters.

Leaving Norfolk 19 August 1950, *Missouri* became the first American battleship to reach Korean waters just one day in advance of the Inchon landings on 15 September 1950. On arrival off Kyushu, Japan, *Missouri* became the flagship of Rear Admiral A.E. Smith, and the next day was bombarding Samchok in a diversionary move coordinated with the Inchon landings.

In company with the cruiser USS *Helena* and two destroyers, she helped prepare the way for the Eighth Army offensive. In a bombardment of the Pohang area 17 September 1950, *Missouri's* 16-inch shells assisted the South Korean troops in the capture of that town and their advance to Yongdok.

Her bombardment of the Mitsubishi Iron Works and the airfield at Chongjin on 12 October were a significant factor in the advance of American and other United Nations forces embattled ashore. Her guns did considerable damage to marshaling yards and a strategic railroad bridge on the Tanchon area. She moved on to bombard Wonsan and then moved into Hungnam 23 December 1950. Her powerful guns hit enemy troop concentrations, command posts, and lines of communication, providing cover for the evacuation of the last of the UN troops from Hungnam on Christmas Eve, 1950. In the opening weeks of 1951, *Missouri* continued coastal bombardment aimed at destroying transportation facilities and disrupting the flow of enemy reinforcements and supplies to central Korea. She joined a heavy bombardment group off Kansong on 29 January 1951 in a simulated amphibious assault which provided a diversion some 50 miles behind the enemies front lines.

During the first week of February, she gave fire support to assist the advance of the 10th US Army Corps in the area of Kangnung. She systematically bombarded transportation facilities and enemy troop concentrations in the vicinity of Tanchon and Songjin. She made similar gun strikes between 14 and 19 March at Kojo Wan, Songjin, Chaho, and Wonsan aimed primarily at transport complexes necessary for the continued reinforcement and supply of enemy forces in central Korea.

Then, on 28 March 1951, *Missouri* was relieved of duty in the Far East and left for the United States and Norfolk, arriving there 27 April 1951. She again joined the Atlantic Fleet to train midshipmen and other prospective naval officers until 18 October 1951 when she entered Norfolk Naval Shipyard for an overhaul which lasted until January 1952.

On 4 August 1952, *Missouri* was again in the Norfolk Naval Shipyard for overhaul being prepared for her second tour of the Korean Combat Zone. She stood out of Hampton Roads 11 September 1952, and by end of October, as flagship of the US Seventh Fleet, she was providing seagoing artillery support to Republic of Korea troops in the Chaho area.

Throughout the remaining months of 1952, *Missouri* was on "Cobra Patrol" along the East Coast of Korea. She participated in a combined air-gun strike at Chongjin on 17 November and

Inspection on the Big "Mo".

Rear Admiral J.L. Holloway and Flag Division, August 1952.

USS Missouri sailors marching in Brazil.

Big Wheels and President in Brazil.

on 8 December was bombarding in the Tanchon-Songjin area. The next day it was Chaho, and 10 December Wonsan felt the power of her guns. During the bombardment of the Hamhung and Hungnam areas *Missouri* lost three of her men when her spotter helicopter crashed into the wintry sea on 21 December 1952. On patrol in early 1953, *Missouri* made repeated gun strikes running swiftly just 25 miles offshore in direct support of troops on land. *Missouri* sustained a grievous casualty 26 March, when her Commanding Officer, Captain Warner R. Edsall suffered a fatal heart attack while conning her through submarine nets at Sasebo, Japan. Her last fighting mission of the Korean War was on 25 March 1953 was to resume "Cobra" patrol where she bombarded the Kojo area.

The *Missouri* was relieved as flagship on 6 April 1953 and left Yokosuka the following day to return to the Atlantic Fleet. She arrived at Norfolk 4 May 1953 and put out almost immediately for a midshipman training cruise to Brazil, Trinidad, Panama, and Cuba. She was back again for overhaul in the Norfolk Naval Shipyard from 20 November 1953 to 2 April 1954. In May, she picked up midshipmen from Annapolis and started a training cruise to Europe. Standing out of Hampton Bays, *Missouri* aligned with the other Iowa class battleships for the one and only time. *Iowa, New Jersey, Missouri* and *Wisconsin* sailed together as the future "Strength for Freedom." *Missouri* visited the ports of Lisbon, Portugal and on 6 June 1954, the Port of Cherbourg, celebrating the 10th anniversary of the Normandy landings or "D-Day." In August she left Norfolk for the West Coast and inactivation. *Missouri* traversed the Panama Canal and made ports of call in Long Beach, San Francisco, and Seattle where tens of thousands of citizens visited the ship. The ship then went to the Bremerton Naval Shipyard for mothballing. There she was decommissioned 26 February 1955 and assigned to the Bremerton Group, US Pacific Reserve Fleet.

Missouri served as headquarters ship of the Bremerton group where she was open year round to visitors. As many as 100,000 people a year visited *Missouri* to see the place on her deck where the Japanese surrendered ending World War II.

New Birth

After almost 30 years at rest, *Missouri*, on 14 May 1984, left her berth in Bremerton and was towed to the Long Beach Naval shipyard for modernization and scheduled recommissioning in June 1986. *Missouri* was recommissioned in San Francisco and departed on an around-the-world shakedown cruise, the first battleship to circumnavigate the world since President Theodore Roosevelt's "Great White Fleet" of 1907-1909. The ship was home ported in Long Beach, California.

In 1987, *Missouri* journeyed to the troubled waters of the Persian Gulf, supporting operations near the Strait of Hormiz. During 1988, *Missouri* participated in the Rim of the Pacific (RimPac) Exercise off the coast of Hawaii. Following a routine shipyard period in early 1989, *Missouri* returned to sea and later in the year participated in Pacific Exercise (PacEx) '89 and visited Pusan, Korea.

Persian Gulf

Missouri deployed to the Persian Gulf in support of Desert Shield. On the first day of Desert Storm, she fired her 16" guns at Iraqi targets inside Kuwait. The USS *Nicholas* (FFG-47) escorted her in and she began shelling targets first. From 4-6 February, she fired 112 16" shells, along with Tomahawk missiles. The ship was finally relieved by the USS *Wisconsin*.

As for the 1990s, *Missouri* is as she was during the 1940s, ready for sea and always ready to answer the call of battle. In November 1993, *Missouri* departed Long Beach for Pearl Harbor, where she was the host ship for the 50th anniversary of the attack on Pearl Harbor. She returned to Long Beach and was decommissioned on 31 March 1992. She was towed to the Bremerton, Washington Shipyard where she has rested as part of the Naval Reserve Fleet.

On 2 September 1995, the USS Missouri Association, Inc. will hold ceremonies at the ship in Bremerton, Washington, honoring those who have served aboard the ship and have passed on, as well as those who have served aboard at the time and are attending the 50th anniversary of the surrender signing.

Upon call, *Missouri* will still be a powerful and fearful dreadnought in the best tradition of the USN.

Postscript: On the 5 January 1995, the Department of the Navy Chief of Naval Operations, by reference of President Clinton and the Board of Inspection and Survey, recommended that the Iowa Class Battleship, including the USS *Missouri*, be stricken from the Naval Vessel Register. This was approved by the Secretary of the Navy John H. Dalton on 12 January 1995 and the ships await their final destiny. (2 September 1995)

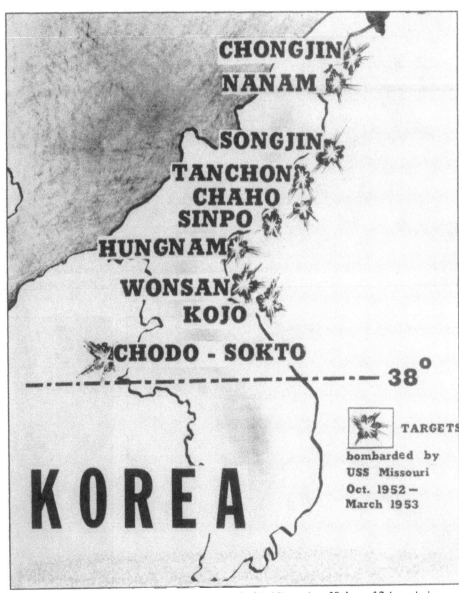

These are the areas shelled by the 58,000-ton battleship Missouri on 38 days of firing missions on her second Korean combat tour. The Missouri shelled Wonsan, the Red's best ice-free harbor, 11 times with heavy bombardments on supply, transportation and gun facilities. At Chongjin, only 55 miles south of the Manchurian border, the Missouri struck three times, the first of which was a coordinated air-sea attack conducted by fast carrier Task Force 77. On her one mission to the Korean west coast at Chodo Sokto, the "Big Mo," which normally was accompanied by a destroyer escort, operated with the British cruiser Birmingham. The ship usually steamed as part of Task Force 77, when she was in the combat but not firing.

Ready to be relieved from Far Eastern duty the battleship Missouri (left) rides from the same buoy as her sistership New Jersey in a Far East replenishing port.

USS Missouri (BB-63) and USS Wisconsin (BB-64) are replenished together by USS Sacramento (AOE-1) in the Persian Gulf during Operation Desert Storm in January 1991.

USS Missouri fires a broadside in 1989 from the dreadnought's nine 16-inch guns.

Former USS Missouri commanding officer Capt. John J. Chernesky tours US Ambassador to South Korea Donald Gregg around the historic battleship during the ship's port call to Pusan, October 1989.

With Missouri in the background, an Iraqi mine bobs in the water, awaiting destruction by the ship's Explosive Ordnance Disposal Unit.

The "Mighty Mo" fires a 16-inch round from gun #2 of turret two during a night fire in Operation Desert Storm.

At left: Robert Webber greeting President George Bush aboard the USS Missouri, December 7, 1991 at Pearl Harbor, HI. **Below:** Aerial view of the homecoming crowd in Long Beach, CA, following Desert Storm, May 13, 1991.

GMG2 Wesley Hancock steadies a 16-inch projectile prior to lowering it into turret one during ammo on load in March 1990 at Seal Beach Naval Weapons station before departing for RIMPAC '90 exercises in Hawaii.

A Tomahawk land attack missile is launched by Missouri and heads toward an Iraqi target during the beginning days of Operation Desert Storm in January of 1991.

Perched high above the ship, EW3 Tom Doyle watches the Mighty Mo enter the Gatun Lock of the Panama Canal.

USS Missouri Special Stories

Signing of Peace Treaty aboard the Mighty Mo

by Lt. Ward R. Munson

Soon after hostilities ended on August 15, 1945, I was assigned to a small task group headed by Commander Byrd to start planning the logistics of the peace-signing ceremony to be aboard BB-63 in Tokyo Bay. For precautionary reasons we followed a Japanese destroyer into the bay where we could see enemy guns and personnel on either side. We wondered, of course, if the Japanese gunners had gotten the word about their country's surrender, but apparently they had for we were not fired upon.

The early morning of September 2, 1945, was damp and cloudy, but not wet enough to alter the detailed plans for the ceremony. One plan however was changed only a day or two earlier. Admiral Halsey, whose flag we were flying, had issued an order that the uniform of the day for officers would be dress white. Admiral Nimitz countermanded that order saying that the uniform for the day would not be dress white, but rather khaki with no ties. "We will meet the enemy in the same uniform we wore to win the war."

At 0700 - 170 newsmen and cameramen boarded the *Mighty Mo* from a destroyer. During the next hour and a half high ranking Army and Navy officials of the Allied powers were taken aboard from destroyers and small craft. Admiral Nimitz came aboard at 0800, and his flag replaced that of Admiral Halsey at the mainmast. General MacArthur came aboard at 0843 and his personal flag was broken along side that of Admiral Nimitz. At about that time, my assignment was to line up all of the U.S. Army officers. "General Stillwell, you will stand here please, Sir. Behind you will be General so and so and so and so. General Doolittle, you will stand here, Sir, and behind you etc." There were not less than 30 high ranking Army officers. After positioning the Army officers, it was I who did the same for the Allied representatives beginning with Admiral Nimitz, followed then by the representatives of the Republic of China, the United Kingdom, the Soviets, Australian, Canadian, French, the Netherlands and New Zealand in that order. It was Ensign Perry who assigned the Navy officers to their places.

This was one high-ranking naval officer who was not there that day and that was Admiral Spruance who was aboard the USS *Iowa* lying off Okinawa. He was waiting there in the unlikely event of a last desperate Japanese attack to sink the *Missouri* and all those aboard. Admiral Spruance would then have assumed command of all US Pacific forces.

The Japanese delegation was not allowed to come aboard until only minutes before the ceremony was to start. When it was announced over the ship's public address system that the Japanese party was coming alongside in a US Destroyer, it seemed that complete silence broke out all over the entire ship. All I remember hearing was the boatswain's whistle piping the delegation aboard and the clump, clump of foreign minister Shigemetsu's wooden leg as he climbed aboard the *Mighty Mo*.

All was now ready for General MacArthur to approach the microphone and open the proceedings by declaring, "It is my earnest hope and indeed the hope of all mankind that from this solemn occasion a better world will emerge out of the blood and carnage of the past." Well, it wasn't a solemn occasion except for the Japanese. For the rest of us, it was a joyous occasion for we would soon be going home to our families and loved ones.

The first to sign the surrender document was Shigemetsu, and under his name in Japanese characters is written "By command and in behalf of the Emperor of Japan and the Japanese government." The next to sign was General Omezu of the Japanese Imperial Command, followed then by General MacArthur who had asked Generals Wainwright and Percival to stand by him as he signed. Wainwright, the US General who had surrendered Singapore, had just been released after surviving three years in prison camps. They were emaciated and haggard. Their new military uniforms just hung on their weary and weakened frames. With the first flourish of his pen, MacArthur turned and handed the pen to General Wainwright. The second pen was given to General Percival. I feel sure that this touching incident brought tears to more eyes than just mine. Then after Admiral Nimitz and all of the foreign Allies had signed, MacArthur closed the ceremony by saying "let us pray that peace now be restored and that God will preserve it always." Thus ended the war with Japan after 1,364 days.

I recall vividly the thoughts that were going through my head at that time. I knew that I was highly privileged to be a witness to this great historic occasion and to have a small part in the structuring of the ceremony. I was no here. I had only done what I was trained to do and certainly no better than others. I felt though that I was there for the thousands and probably millions who were more deserving than I. Having been in communications, I quickly thought of names and places familiar to me that raced through my mind in rapid succession: Pearl Harbor, Guadalcanal, Saipan, Tinian, Eniwetok, the Solomon Islands, Iwo Jima and Okinawa. These were places where battles were fought and won that we might be where we were. These were the places where Navy men went to the bottom of the sea with their ships, where Army and Marine personnel died on the beaches and in fox holes and where American and Allied pilots were shot out of the sky. It was for all of these valiant fighting men, the real heroes, who could not be here that we were standing in.

But my personal thoughts were suddenly shattered by a tremendous roar as 1900 Allied planes swooped over the decks of the *Missouri* in a tremendous show of power. Wave after wave came, each a little lower than the one before until the last one barely cleared our stacks.

It was at that very moment it seemed that the sun broke through the clouds for the first time that day, shining off the fuselages of the planes overhead and making visible the peak of Mt. Fuji. The thought came to my mind and probably to many others - "God is now looking down, giving His blessing to this historic occasion."

This I know: I shall never forget the drama, the excitement and the joy of that day, September 2, 1945 aboard the *Mighty Mo* in Tokyo Bay, Japan.

Air Battle Over Task Force 77

by Elwood E. Alexanderson

Documented by "The Sea War In Korea," Malcom W. Cagle, Commander USN; Frank A. Mason, Commander USN; US Naval Institute, Annapolis, Maryland.

It was a cold, gray afternoon in the northern reaches of the Sea of Japan. The date was November 18, 1952. The sea was relatively calm, only small white-caps indicating the wave action. The USS *Missouri* was again sailing with Task Force 77, having only rejoined the group the previous evening following a gun-strike against Chonjin, a main industrial area only 90 miles south of Vladivostok, USSR. The local time was approximately 1330. Noon chow was over and the crew had just turned-to for the afternoon. I was working in the Main Frame for Mk 25 Radar, Sky 3, starboard side on the 03 level amidships.

At approximately 1335, the general alarm sounded, followed by the boatswain's-mate-of-the-watch passing of the word, "General Quarters, All Hands Man Your Battle Stations."

Surrender signing.

Bob Lathrop, Claude Quick, West Riemer, and Alvin Getzs in Honolulu, 1945.

I particularly remember, he did not say, "Drill General Quarters." Since General Quarters for our gun-strikes were announced ahead of time in the Plan of the Day, this call to General Quarters was unusual. When I got to my GQ Station in Air Defense Aft, everyone wanted to know what was going on. An officer, already manning the sound-powered phones, said CIC had reported unidentified aircraft. Since this wasn't the first time that unidentified aircraft had been reported, we didn't think to much about it. We didn't hear anything more until the officer said word had been passed to "load and cock the 40mms." That got our attention for awhile, but time went on and nothing else happened. After about 30 minutes, the word was passed to "Secure from General Quarters."

Approximately eight years passed before I happened to read about this incident in the book, *The Sea War In Korea*. It turns out this action was one of the most dramatic incidents of the Korean War.

The unidentified aircraft turned out to be seven Russian MiG-15s headed straight for the fleet. A Combat Air Patrol (CAP) composed of four F9F5 Panther aircraft from the USS *Oriskany's* VF-781 operating at 13,000 feet above the Task Force was vectored to climb toward the unidentified aircraft. Because of a fuel-boost pump problem in one of the Panthers, only three climbed to meet the MiGs.

An extremely mixed up air action took place between the three Panthers and seven MiGs in which the Oriskany pilots were officially credited with two MiGs destroyed and one damaged. Later compilations by radar and pilot interviews indicated a strong possibility that only one or two of the original seven MiGs returned to base. Five or perhaps even six were either shot down directly, damaged so severely as to crash, or ran out of fuel on the way home. One of the MiG pilots parachuted into the sea and his position was reported by one of the Panther pilots, but no rescue of the pilot was achieved by surface vessels.

It so happened that President-elect Dwight Eisenhower was visiting in Seoul, Korea at Eighth Army Headquarters at the time and Vice Admiral J.J. Clark, Commander Seventh Fleet aboard the *Missouri*, had been invited by General Van Fleet, Commander Eighth Army, to come and meet the next president. Vice Admiral Clark delightfully accepted the invitation and took the three *Oriskany* pilots with him. The pilots got to meet Eisenhower and related their stories to him. The pilots concluded that they were fortunate to have come back with their whole skins. Vice Admiral Clark, in the presence of the next President of the United States and Commanders of the major Army and Air Force units in Korea, felt that the Navy had definitely "stolen the show."

One of the biggest Navy air battles of the Korean War occurred 35-40 miles from us and it took over eight years before I read in a book what had actually happened.

Aboard the Battleship *Missouri*

by Quentin Beaty

We had formed a small string band on the ship. There were five members all from different states. The only name I can remember was a fellow from California named Mitchell. Mitchell had picked up a pretty good guitar with him on his battle station, which was just below my battle station on 40mm quads on the portside. This day an air raid was taking place and Mitchell had to lay this guitar down to man his battle station and this Japanese suicide plane was trying to crash on the ship. We had him shot up pretty bad and as he came over our ship, a piece of the plane fell on Mitchell's guitar and broke it up pretty bad; I can still remember how mad he was at the Japanese.

By the way, if any of the members of that string band read this, drop me a line. I would like to hear from you. I was the fiddle player in the band and now live in Muncie, Indiana.

The Other Three USS *Missouri* Ships

by D.W. Bishop Jr.

The *Mighty Mo* that is soon to be recommissioned is actually the fourth ship bearing the name USS *Missouri*. The first three mirror the history of American seapower for the better part of a century, from wooden ships and the early days of steam forward to the present era.

The first *Missouri* was one of the first steam-powered warships of the United States Navy. A great wooden side-wheel frigate, she was built in the New York Navy Yard in 1842. She was armed with two 10-inch and eight 8-inch shell guns (larger than most guns today), had an overall length of 229 feet, and could spread 19,000 square feet of canvas. Her paddle wheels were 28 feet in diameter and 11 feet wide.

After a prolonged cruise in the Gulf of Mexico, the first USS *Missouri* was ordered to return to Washington in the Spring of 1843 to prepare for foreign service. She sailed in August for the Mediterranean with Caleb Cushing, the

U.S. Minister to China, embarked, bound for Alexandia, Egypt, on the first leg of a journey to conclude negotiations for a commercial treaty with China. After 19 days at sea, the frigate reached Gibralter. Alas, the next day, August 26, 1843, a fire ignited in a storeroom and spread so rapidly that all hope of saving the ship had to be abandoned, and her crew barely escaped with their lives. In a few hours, the splendid frigate was reduced to a blackened and sinking hull in the harbor.

The second USS *Missouri* lasted little longer. Originally a Confederate iron-clad ram, she was launched at Shrevesport, LA, April 14, 1863, as a center wheel steam sloop, having a length of 183 feet and a beam of 53 feet.

Her principal Confederate service was transport and mining details between Alexandria and Shrevesport above the Red River obstructions. At the close of the Civil War, she returned to Shrevesport and surrendered to Union Forces on June 3, 1865. She became a part of the USN, although she was never activated as a commissioned ship. This service was brief, and she was taken to Mound City, Illinois where she was sold for scrap November 29, 1865.

Somewhat more glorious a fate occurred to the third USS *Missouri* (BB-11), built in Newport News, Virginia, and commissioned December 1, 1903. The battleship, almost 400 feet long and weighing more than 12,000 tons, had armament that included four 12-inch .45 caliber and 16 6-inch .50 caliber guns, and had a speed of 18 knots.

Her first four years consisted of cruises to the Caribbean and Mediterranean. She rendered aid to earthquake victims at Kingston, Jamica and participated in the Jamestown Exposition. The battleship was one of 16 dreadnaughts in the "Great White Fleet" that passed in review before President Theodore Roosevelt at Hampton Roads, Virginia, December 16, 1907, as they began the first leg of their world cruise. The mighty warships made numerous port visits as they passed around South America to San Francisco; in each port there was tremendous enthusiasm and acclaim for the American fleet. From San Francisco the Great White Fleet steamed to

R. Division, 1945: (top) Pajak, Richardson, Morris, (bottom) Dell, and Fredrikson.

USS Missouri, May 24, 1951.

Modern day sailors uncover the ship's bell from the 40s prior to pulling into Pusan, South Korea, October 1989.

Hawaii, New Zealand, and Australia, where they were greeted by wildly cheering crowds. The ships continued to Manila and Tokyo Bay, where the Americans were greeted by thousands of Japanese who had learned to say "We are glad to see you" in English. At Amoy, China, six great pavilions had been specially built to host a banquet and entertain the sailors. The fleet then transited the Suez Canal to reach Port Said, Egypt. From there, *Missouri* proceeded with the battleship *Ohio* to visit ports in Greece and Turkey. President Roosevelt greeted the Great White Fleet as it returned to Hampton Roads February 22, 1909. The world cruise had brought prestige and enhanced goodwill toward the United States and its Navy.

During the next seven years, the battleship was placed in reserve three times; each time to be recommissioned to serve with the Naval Academy Practice Squadron for midshipman cruises in the Mediterranean and Pacific. When the United States entered World War I, *Missouri* joined the Atlantic Fleet, throughout the entire war operating in the Chesapeake Bay and along the eastern seaboard training thousands of recruits. After the Armistice, she made four trips to France to transport returning doughboys. She was finally decommissioned for the last time September 9, 1919, and four months later met the ultimate fate that grieve sailors most for the ships they love - she was sold for scrap under terms of the Washington Treaty which limited naval armaments.

While the first three *Missouris* served honorably and well for varying periods of our nation's history, it remained for the fourth and

current USS *Missouri* to achieve a name and a fame that ranks with the most famous vessels in all recorded history. And she will soon be ploughing the seas again.

History Of The USS *Missouri*

by D.W. Bishop Jr.

On January 6, 1941, 11 months before Pearl Harbor, the keel was laid for the last battleship ever built by the United States. The fourth ship named for the Show-Me State, USS *Missouri* (BB-63) was the most modern of the four Iowa class battleships, the mightiest gun platforms ever built by this nation.

Two of *Missouri's* sister ships, *New Jersey* (BB-64) and *Iowa* (BB-61), are now back in the fleet, and the other, *Wisconsin* (BB-62) will rejoin the fleet in 1989. *New Jersey*, which saw duty in Vietnam, was recommissioned for the third time in December 1982 and *Iowa* for the third time in April 1984. Wisconsin's recommissioning will be her third, while it will be the second for the *Missouri*.

The great battleship, unsurpassed in size, firepower and maneuverability, was built at a cost of $92 million at the New York Naval Shipyard. The majestic ship is 887 feet long and the equivalent of an 18-story building in height, seven being under water. Its deck space covers 418,000 square feet. It was christened at the Brooklyn Yard on January 29, 1944, by her sponsor, Margaret Truman, daughter of the then-senior senator from Missouri, Harry S. Truman. There were 26,000 shivering persons at the launching who heard Truman, the principal speaker, speak of

the ship's "blazing batteries" and make the prophetic statement that "the time is surely coming when the people of Missouri can thrill with pride as the *Missouri* and her sister ships sail into Tokyo Bay."

Little could anyone present know that both the ship and Truman soon would be central figures in history; the battleship leading the fleet to the final victory over Japan, and the unassuming senator leading the nation and the free world to that victory.

USS *Missouri* was commissioned on June 11, 1944. Following the usual shakedown period and routine check-ups, the *Missouri*, under the command of Navy Captain William M. Callaghan, got underway from the West Coast and arrived in the forward area of the Pacific in early February 1945, in time to support the Allied landings on Iwo Jima and Okinawa and to escort Task Force 58 in the first mass air strikes against Japan itself. Her massive 16-inch guns took part in the first naval bombardment of the Japanese home islands. The ship won three Battle Stars, survived two kamikaze hits with only a few dents, and was the flagship for Admiral William F. Halsey Jr., Third Fleet commander, in the final months of the war.

The signing of the surrender documents ending World War II occurred on the teak deck of USS *Missouri*, making the ship one of the most famous in all history (the ship was designated a national historic landmark in 1972).

With a Missourian in the White House whose daughter was the ship's sponsor and who was a patron of the battlewagon, it is not surprising that the *Missouri* avoided decommissioning, unlike her sister ships, once the war ended.

Most of the interlude between World War II and Korea was routine. In March 1946, President Truman ordered the *Missouri* to carry home the body of the Turkish ambassador to the United States (he had died during the war). The battleship's visit was a gesture of support for the independence of Turkey and its neighbor Greece at a time when both nations were seriously threatened by Soviet expansion. Turkey commemorated the ship's visit with issues of postage stamps. If this was the high point of the peace years, the low point occurred one unfortunate day in 1950 when the ship suffered one of the worst gaffes possible; it ran aground within public view near Norfolk, Virginia. This national humiliation led to courtmartials, reduction in crew, and assignment as a permanent floating classroom.

But *Missouri* restored her honor when the Korean War broke out, ironically, another grounding, this one deliberate, helped her do it. When the war broke out, the *Mighty Mo* was the only battleship then in commission, and it quickly steamed 10,000 miles to be in the forefront of American forces called to Korea. During her first deployment she supported the United Nations' landing at Inchon Harbor and assisted the American evacuation of Hungam.

When the Chinese Communists entered the war and started pushing UN forces back toward the ocean, the *Missouri* was called in for gunfire support over the heads of the troops to give them space to retreat. The task force commander ordered the ship into shallow water, but the ship's skipper noted that the ship would go around. It was agreed that the *Missouri* would be deliberately grounded to provide a seven-degree list at low tide, increasing the range of the guns five miles and adding support for the evacuation. When the tide came in again, the ship refloated and went back out to sea.

In eight months of duty and two deployments the *Missouri* bombarded Communist positions and provided gunfire support to UN troops all along the Korean coast, earning five Battle Stars, two more than in World War II.

With mentor Truman out of office and the war over, the giant vessel was decommissioned in February 1955 and taken to the Puget Sound Naval Shipyard in Bremerton, Washington, for mothballing and viewing as a tourist attraction. Truman complained that his favorite ship was hidden in a "closet, where nobody can see it."

But an average of 180,000 people did visit the ship each year for the almost 30 years that the *Missouri* was berthed at Bremerton. The ship was not allowed to rust away during these three decades; it was preserved in almost time-capsule condition. Within the eight sealed airtight zones, no spark hammers and wrenches used around gunpowder stood ready for inspection. Brass fitting still shone. Waiting barber chairs, work orders of the day, and inventory sheets were found open as if the crew were returning momentarily.

All the compartments, the hull, shafts, screws, prop, rudder, and all other increments of the ship were found to be in top condition. The sealing and dehumidifying of the ship kept her cybernating like an 11-year-old ship (1944-55) rather than the chronological 40-year-old that she was when she was taken out of mothballs in 1984.

Missouri was towed to the Long Beach Naval Shipyard in May 1984 to begin her reactivation process as the third Iowa Class battleship to be recommissioned.

Work is nearing the final stages on the reactivation and refitting of the ship to make it thoroughly state-of-the-art in armament, equipment and capability. The Stars and Stripes and the Union Jack were hoisted on *Missouri* on September 12 for the first time in 30 years, marking an important milestone in the overhaul and modernization. On that day Boiler Technician Third Class Gene Nowakowski of Jefferson City, Missouri became the first crew member to be berthed aboard ship. Another important milestone occurs in December 1985 with the first operational sea trials.

USS *Missouri* is scheduled for recommissioning on May 10, 1986. She will be homeported in San Francisco with her crew of some 1,500 (representing a $60 million annual payroll).

In a 1949 Navy Day address, President Truman assessed the role of the ship, and his words still seem appropriate today: "The *Missouri* is not a symbol of might in war," the President observed, "but a symbol of might in peace."

On Board The USS Missouri

by Donald H. Bloch Sr.

The weather and sea conditions around Korea in the winter months are unpredictable. Ask any aerographer aboard the aircraft carriers that were there. I mention this because as you will see, it brought about a testing experience for me.

One of my duties as a quartermaster second class on the *Missouri* was to take the helm when special steering skills were required. One day, while steaming off Korea, we were to approach and go alongside a Navy oiler to take on fuel while underway. When the Bos'n mate passed the word to "man all fueling stations," that meant, for me, to take over the helm.

Having relieved the regular helmsman, we preceded to approach the oilers portside. We went alongside and kept station on the oiler while all the tag lines were passed over along with the

fuel lines. When all lines were aboard and hooked up, the oiler commenced pumping oil to us. This is not a short time situation.

While we were alongside for a length of time, as usual, the weather had been changing, especially the wind direction, which was changing the direction of the sea.

The oiler was experiencing difficulty in maintaining a steady course because of the sea coming in about 10° off her starboard bow and our wash against her portside.

Captain Duke and the skipper of the oiler talked the situation over, and came to the conclusion to change course 10° to the starboard, while hooked up and pumping oil. This would bring the wind and sea in from dead ahead making steering control more manageable.

This meant that the turn had to be made in unison while maintaining distance between ships and position fore and aft.

It was decided to change course in five increments of 2° each. We would swing in unison 2° to starboard then steady up, then swing again 2° to starboard and steady up and repeat until the 10° course change was complete. The oiler executed the turns while passing the word to us at the same time. We then executed our turn.

I had learned the way the *Missouri* handled and answered to the wheel, while standing condition three watches on the helm. A good helmsman must be able to correctly anticipate the movements of his ship.

The *Missouri* handled extremely well during the maneuver and everything went well on the new course. I received a "well done" from the captain and operations officer who, incidentally, was the officer of the deck and had the conn during this maneuver. To this day I feel proud of my accomplishments while on board the *Missouri*. The tour of duty was one big memorable event in my life.

Lucky Strike

by Michael F.X. Boylan

In the early 1940s during World War II, the American Tobacco Company changed the package design of their very popular Lucky Strike cigarettes. The green package with the red circle

KE Division (Radar Repair), October, 1945. Top:(left to right) Lourenco, Boulay, Mason, Boylan, Katz, Enten, Nason, Schmitt, DeCarlo, Roberts. Middle: Kennedy, Riddle, Freeman, Carmint, Shahan, Williams, Brackin (Commanding Officer), Coccione, Nadaw, Zellers. Bottom: McCloskey, Foat, Woods, Irwin, Banks, Smith, DiBella, Haley, Hammond, Vallario.

General view aboard the Missouri during the initiation ceremonies that made Shellbacks of some 1,400 Pollywogs, including the Trumans.

The President and his daughter enjoy the antics.

on the front was changed to a white package with a red circle. The Madison Avenue hucksters came up with the slogan "Lucky Strike green has gone to war." This slogan was widely disseminated through billboards, magazine ads and radio commercials. The connotation was that in switching from green to white, valuable copper, used in the manufacture of green ink, was being "salvaged" for the war effort.

A few days before the USS *Missouri* entered Tokyo Bay for the September 2, 1945 peace signing and while we were still outside the bay, I observed a small boat coming alongside the *"Mighty Mo."* Several Japanese harbor pilots came aboard carrying many navigation charts. These charts identified the location of ship mines and obstacles in Tokyo Bay. Without this information no ship could safely enter the harbor. Accompanying the harbor pilots was a military escort of three or four enlisted men (I couldn't identify them as either soldiers or navy personnel). These men were armed with small sidearms (revolvers). As they came aboard they were disarmed by the on-board fleet Marines who, anticipating the arrival of the Japanese entourage, were assigned to the gangway. The Japanese enlisted men were not allowed to accompany the harbor pilots to Admiral Bull Halsey's wardroom, but were detained by the Marines at the gangway on the main deck of the ship.

After the Japanese harbor pilots had concluded their business with the Admiral's staff, they and the Japanese enlisted men returned to Tokyo by small boat. I later learned from one of the marines, who was at the gangway, that when one of the marines took out a pack of Lucky Strikes and offered a cigarette to a Japanese enlisted man, the Japanese asked in broken English "What happen to green (on the package)." The

marine was perplexed and hesitant to state the slogan "Lucky Strike green has gone to war." He told me he accordingly mumbled some inaudible reply and let it go at that.

Presidential Cruise

by George W. "Bill" Butch

Rio de Janeiro was an exciting place to visit, a lot of things to see and some nice things to buy especially butterfly wing trays which I have to this day. The big thing about this trip was the return voyage home.

It was August 30, 1947 when President Truman, his wife Bess and daughter Margaret were returning from Rio on a leisurely two week trip back to the States. They were accompanied by Secret Service, the White House physician, several aides and a mob of photographers.

We would see President Truman every morning while we were in the chow line. He was taking his morning walk which meant several trips around the main deck. About every 10 feet, he would say, "Good morning, good morning."

They had omitted the shellback ceremony on the way to Rio and held it on the return trip so that they could "share" it with the President. Thousands of dollars worth of theater costumes were rented and weeks of preparations were made by the shellbacks.

Summons were issued to all pollywogs, including the President, and on the eve of the big day things were beginning to look bad. The big day finally came and I knew I was in for it when I saw all those signs on the main deck, several pirates and other characters pulling wag-

ons and carts and taking prisoners. Their banner said, "Expect no justice." My breakfast was navy beans, salt crackers and water.

The Secret Service were some of the first to be initiated. The President did not escape either, he was sentenced to give a "Republican speech." His daughter Margaret had to sing a song and his wife Bess just laughed and let the charges pile up.

Later in the day I was standing on the fantail watching the activities wind down. I was dressed in the uniform of the day, rolled up dungarees, tee shirt, barefoot and hatless, and paddling my foot in a puddle of water when I happened to look behind me, and there he was - the President of the United States. I glanced again and he was gone. This was an interesting two weeks aboard the *Missouri*. It could never happen again.

A Few Fighting Days Aboard
The USS Missouri

by Donald Burr

The image on the ship's radar flickered. It twinkled, disappeared for a moment, then came back. We had a contact on something that was to small to be a ship. It always came back into view at the same point, so we knew it was real; not a figment of the radar's electronic imagination. The Combat Information Center officer looked at it for a few moments, then reached for the phone to the bridge. "Captain, CIC. We have a definite surface contact on the radar. Range is four point two miles, bearing 086 degrees. It's too small for a surface ship. It may be a submarine on the surface."

"Very well. I will report it to the admiral." Captain William Callaghan of the battleship USS *Missouri* radioed the contact report to Rear Admiral Arthur Radford, who was on his flagship aircraft carrier a mile on our beam. After several minutes the word came from the admiral's staff, "Your radar must have a glitch. None of other ship's radar's can see his target."

Topside, the weather was almost perfect for a seagoing cruise. The air was mostly clear and quite brisk. The sky was blue and speckled with a few clouds. The deep blue sea hissed and sloshed endlessly past the speeding ships. One thing was wrong. This was March 23, 1945. We were not, I repeat not, on a pleasure cruise.

All these ships were here in the western Pacific Ocean to force the end of a long and viciously fought war against the Empire of Japan.

The *"Mighty Mo"* was part of fast carrier Task Force 58, commanded by Admiral Marc Mitscher. It was the mightiest fighting force ever assembled. *Missouri* steamed in Task Group 58.4, one of five task groups that made up TF-58. This group included the aircraft carriers: *Yorktown, Intrepid, Enterprise* and *Langley*. Close by the carriers to provide a concentrated barrier of antiaircraft fire for protection against enemy planes were the *Missouri,* the battleship *Wisconsin,* and cruisers: *Alaska, Guam* and *San Diego*. About four to five miles out from the tight grouping of major warships was a loose ring of

destroyers. Their chore was to intercept any Japanese submarines who might come trying to sink a warship of the USN. The destroyers also were able to report any Japanese planes flying too close to the water for radar detection. The task groups were separated by about 30 miles, but all were launching air attacks against the main enemy bases on the home islands of Japan.

The radar operators and officers in the *Missouri* CIC continued to watch the image on the radar scope. It disappeared for short periods, but always came back in the same location. It was definitely not a radar glitch. As the blowers hummed softly and the ship vibrated faintly from the thrust of the ship's mighty engines, we kept the captain informed of the contact. He decided to report the contact to the admiral again. This time, Admiral Radford (later to be head of the Joint Chiefs of Staff of the United States) replied that what we had picked up was one of our own destroyers.

Missouri's navigator checked the plan of the day, and the standing orders for ship assignments. The radar contact was definitely in the right location for a destroyer but this day no destroyer was assigned to that post. The task group did not have enough destroyers to fill all the possible locations, so that position was not occupied by one of our ships.

After a short huddled conference between the captain and the navigator, this was called to

the admiral's attention. He sent a destroyer boiling across the sea at flank speed to check it out. We, in the CIC, watched the radar "blip" disappear as the destroyer got near it. Later, we heard that the destroyers had sunk the sub.

Lieutenant Ronald Loveredge eyed the radar screen, "Hot damn, that really was a sub." CIC smiled, "Did you think it wasn't?"

I was very proud that day. One of "my" radars was the best in the fleet! (Of course, it helped that our SG-3 radar was mounted 165 feet above the waterline, higher than any other except for the battleship *Wisconsin*). *Wisconsin* was built from the same blueprints as *Missouri*, but I don't know what their radar equipment was.

The best weapon the Japanese had at that time, near the end of World War II, was their kamikaze force. They had lost hundreds of their most highly trained air crews three years before at the battle of Midway. US forces sank four of their best battle carriers, and the pilots in the air had no place to go but into the sea. The Japanese never were able to replace these losses.

Several centuries before, the islands of Japan were attacked by a huge fleet of ships from China. Japan seemed doomed. At the last moment, a terrific storm came up and destroyed the attacking ships. The Japanese called this storm a "Divine Wind" or kamikaze.

Now they intended to create their own Divine Wind to drive off the invading American

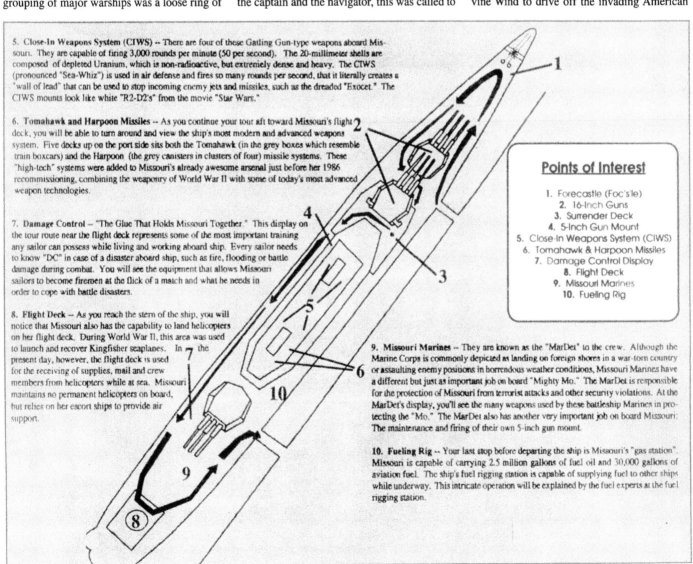

5. **Close-In Weapons System (CIWS)** -- There are four of these Gatling Gun-type weapons aboard Missouri. They are capable of firing 3,000 rounds per minute (50 per second). The 20-millimeter shells are composed of depleted Uranium, which is non-radioactive, but extremely dense and heavy. The CIWS (pronounced "Sea-Whiz") is used in air defense and fires so many rounds per second, that it literally creates a "wall of lead" that can be used to stop incoming enemy jets and missiles, such as the dreaded "Exocet." The CIWS mounts look like white "R2-D2's" from the movie "Star Wars."

6. **Tomahawk and Harpoon Missiles** -- As you continue your tour aft toward Missouri's flight deck, you will be able to turn around and view the ship's most modern and advanced weapons system. Five decks up on the port side sits both the Tomahawk (in the grey boxes which resemble train boxcars) and the Harpoon (the grey canisters in clusters of four) missile systems. These "high-tech" systems were added to Missouri's already awesome arsenal just before her 1986 recommissioning, combining the weaponry of World War II with some of today's most advanced weapon technologies.

7. **Damage Control** -- "The Glue That Holds Missouri Together." This display on the tour route near the flight deck represents some of the most important training any sailor can possess while living and working aboard ship. Every sailor needs to know "DC" in case of a disaster aboard ship, such as fire, flooding or battle damage during combat. You will see the equipment that allows Missouri sailors to become firemen at the flick of a match and what he needs in order to cope with battle disasters.

8. **Flight Deck** -- As you reach the stern of the ship, you will notice that Missouri also has the capability to land helicopters on her flight deck. During World War II, this area was used to launch and recover Kingfisher seaplanes. In the present day, however, the flight deck is used for the receiving of supplies, mail and crew members from helicopters while at sea. Missouri maintains no permanent helicopters on board, but relies on her escort ships to provide air support.

Points of Interest

1. Forecastle (Foc's'le)
2. 16-Inch Guns
3. Surrender Deck
4. 5-Inch Gun Mount
5. Close-In Weapons System (CIWS)
6. Tomahawk & Harpoon Missiles
7. Damage Control Display
8. Flight Deck
9. Missouri Marines
10. Fueling Rig

9. **Missouri Marines** -- They are known as the "MarDet" to the crew. Although the Marine Corps is commonly depicted as landing on foreign shores in a war-torn country or assaulting enemy positions in horrendous weather conditions, Missouri Marines have a different but just as important job on board "Mighty Mo." The MarDet is responsible for the protection of Missouri from terrorist attacks and other security violations. At the MarDet's display, you'll see the many weapons used by these battleship Marines in protecting the "Mo." The MarDet also has another very important job on board Missouri: The maintenance and firing of their own 5-inch gun mount.

10. **Fueling Rig** -- Your last stop before departing the ship is Missouri's "gas station". Missouri is capable of carrying 2.5 million gallons of fuel oil and 30,000 gallons of aviation fuel. The ship's fuel rigging station is capable of supplying fuel to other ships while underway. This intricate operation will be explained by the fuel experts at the fuel rigging station.

USS MISSOURI

Vital Statistics

Built	New York Naval Shipyard, Brooklyn, NY
Class	IOWA-Class Battleship
Keel Laid	6 January 1941
Launched	29 January 1944
Commissioned	11 June 1944
Operational	14 December 1944
Decommissioned	26 February 1955
Recommissioned	10 May 1986
Length	887'3" (270.4 m)
Beam	108'3" (33 m)
Draft	38' (11.6 m)
Height	209'8" (63.9 m) from keel to mast top
Displacement	58,000 tons (53,000.000 kg) fully loaded; 45,000 tons (41,000,000 kg) unloaded
Personnel (WWII)	134 officers, 2,400 enlisted
(Today)	1,515 Navy - 65 officers, 1,450 enlisted
	53 Marines - 2 officers, 51 enlisted
Boilers	Eight 600 Pounds per Square Inch (413.69 Newtons per Square Centimeter) Babcock & Wilcox
Main Engines	Four geared General Electric turbines
Horsepower	212,000 shaft horsepower (total of all 4 shafts)
Propellors	Two five-bladed 17'5" (5.3 m) (inboard)
	Two four-bladed 18'3" (5.6 m) (outboard)
Rudders	Two
Speed	In excess of 30 knots
Tank Capacity	2.5 million gallons (9.5 million liters) of fuel oil
	30,000 gallons (114,000 liters) of aviation fuel
	239,000 gallons (905,000 liters) of fresh water
Armor	The main armor of the hull is 13.5" (34.29 cm) thick. Other armor thicknesses are: Turret faces - 17" (43.2 cm); Turret tops - 7.25" (18.4 cm); Turret backs - 12" (30.5 cm); Turret sides - 9.25: (23.5 cm); Second deck armor - 6" (15.2 cm) and Conning tower sides - 17.3" (43.9 cm).

Armament

Main Gun Battery	Nine 16"/50 (406 mm) caliber guns in three, three-gun turrets
	Range: 23 miles (42 km)
	Projectiles: Amor Piercing - 2,700 pounds (1,225 kg)
	High Capacity - 1,900 pounds (862 kg)
	Powder: Standard load of six 110 lb. (49.9 kg) bags
	Rate of Fire: Two rounds per minute, per gun
	Use: Anti-surface, shore bombardment
Secondary Gun Battery	Twelve 5"/38 (127 mm) caliber guns in six dual-gun mounts
	Range: 9 miles (16.5 km)
	Projectiles: 55 lbs. (24.9 kg)
	Powder: 30 lbs. (13.6 kg)
	Rate of Fire: 15 rounds per minute, per gun
	Use: Anti-surface, anti-air, shore bombardment
Air Defense:	Four Phalanx Close-In Weapons Systems Gatling-style guns capable of firing 20mm ammunition at a rate of 50 rounds per second for self-defense against missiles and aircraft.
Missiles:	Capable of carrying the Tomahawk Anti-Surface Missile/Land Attack Missile and the Harpoon Anti-Surface Missile.

Our main fleet had three lines of defense against these determined and brave warriors. (At the time no one in our beleaguered fighting force thought of them as anything but foolish and deadly enemies). First defense was our far ranging swift carrier launched fighter planes. They were guided by radar controllers on the carriers. They sent down to flaming death most of the kamikaze aircraft before they got within sight of our vigilant sailors on lookout duty.

Those attackers who got past the fighter planes were sighted by our fire control radar at a distance of several miles. They came under fire from the 5-inch guns on every ship of the fleet. The white puffs of bursting high explosive shells marched across the sky in search of a target. *Missouri's* 5-inch guns splashed a number of enemy aircraft during this period.

We learned to cringe a little when we heard our 40mm and 20mm machine guns start to fire. This meant that a Japanese attacker had made it through the 5-inch barrage, and was within a mile of the ship. These weapons were aimed mostly by sighting along the streams of smoke from tracer shells included in their ammunition.

On April 11 the greatest wave of determined death and destruction was sent against our carrier group. Some of these young men were not trained well enough in navigation to find a target. They went down into the cruel, empty ocean as complete failures. Thirteen of them were sent down in flames by the alert patrolling fighter planes or our five inch guns. Three survived long enough to streak in close to the water and find our concentrated group of valuable fighting ships.

One of them closed in on *Missouri* from astern. He was bathed in a shower of twisting, gyrating white tracer smoke as our gunners reached out to touch him. He banked and turned in vain. Small puffs of smoke and bits of aluminum sparked off his darting plane, but none of the hits stopped this determined pilot. His bomb was knocked off into the water. Still he came on into our hail of machine gun fire. The loss of his bomb gave him more lift, and he smashed his battered plane into *Missouri* at the edge of the main deck. Machine gunners to his right and left marveled at their luck in not being in his path. Our ship's photographer got a clear picture of this plane a split second before the impact.

I was at my battle station in the Combat Information Center below decks, linked to all my radar technicians by a party line telephone system. Each of them stood guard at a transmitter of the search radar of fire control radar systems. Some of them were in stuffy little cubbyholes with nothing but the humming transmitter to watch. First class radar technician, Donal Nason, was assigned to a fire control radar compartment welded to the side of the forward stack. He stepped out onto a catwalk high above the deck, and was relating the action to those of us locked inside. A wing partially filled with volatile aviation fuel flew off the crashing kamikaze, and came to rest on the upper deck. It narrowly missed landing on top of one of our 5-inch gun mounts. Unfortunately, it burned directly in front of a large air intake. Giant blow-

forces. As the end neared for the Japanese Empire, they enlisted hundreds of volunteers for suicide duty. The young teenage pilots were given an absolute minimum of flight training. They all swore to die to protect their God, the Emperor from the evil American forces closing in on Japan. Any plane which could be made to fly was loaded with high explosives and just enough fuel for a one-way flight. These kamikaze warriors then staggered into the air and were pointed in the direction of the U.S. ships. When one of them saw an enemy (the bigger, the better), he put his plane into a screaming full power dive and crashed into the target. They were extremely hard for the American gunners to hit because of the high speed approach.

The most vulnerable ships were our destroyers, assigned to early warning picket duty. They were operating alone without supporting antiaircraft fire. Their thin hulls could not withstand even the crash of a kamikaze without a bomb. Losses were heavy.

ers blasted a torrent of fresh air into the forward boiler room. There was a little panic in the boiler room as thick clouds of smoke and gas fumes poured in, but no one was hurt.

The gasoline fire was almost under the platform where technician Nason was giving a play by play account of the action. I heard him drop his phone and leave in a hurry. Moments later, he checked in from the next radar compartment.

Total damage to the *Mighty Mo* was some burnt paint and some shattered nerves. Next day, the kamikaze pilot was buried at sea with military honors. The wing of the plane was turned over to the sailors to be cut up for souvenirs. Most prized were bits of the aluminum with Japanese inspector's stamps on them. I got a magnesium fitting which I filed into a ring. The pilot was buried at sea the next day. (Members of the crew refused to stand at attention during the ceremony, and many of them said he shouldn't have a Christian burial.) A very angry Captain Callaghan got on the bullhorn and said, "This is not a Christian burial! It is a military ceremony, as required by the Articles for the Government of the USN." The crew members complied.

A few days later, another major Japanese air attack raged. For about 12 hours the fighter planes from the carriers kept the incoming forces away from out fighting ships. Finally, one group of kamikaze planes got through in the afternoon. Our 40mm machine guns with their chunka-chunka-chunka rhythm downed a Zero type attacker trying to crash out his life against the carrier *Intrepid*. Two minutes later another Zero spun out of control over the *Missouri*. He clipped the top of our aircraft crane on the stern, narrowly missed two batteries of 40mm guns, and slammed into the water near our rudders. His bomb went off and threw the *Missouri* around a little. I was going down a ladder on the way to my battle station, and almost fell down the hatch.

In the ammunition handling rooms below the 40mm mounts, clips of live ammunition were bounced around the compartment. Luckily, none of it went off. Two chief petty officers, lulled by many false alarms, stopped in the CPO head to relieve themselves before going to General Quarters. When the bomb exploded, many of the toilet bowls and urinals were shattered. It was reported that one of the chiefs got halfway through the next compartment before he got his stream turned off.

Damage Control reported that one of our twin rudders might be bent. This could cripple our speed and maneuvering ability. Fortunately the rudder was not seriously damaged.

Thus ended the damage to the *Missouri*, although more kamikaze attacks continued against Task Force 58 until the war ended.

NOTES: These events happened. Based on my memories 50 years later, they are correct. I am indebted to the book, *Mighty Mo* by Gordon Newell and Allan E. Smith for filling in the names of the ships involved in Task Group 58.4 and some of the dates. Names (ones that I remember) used in this tale are real people.

Sweet Mother

by Chad H. Collins, USS Missouri

Sleep sweet Mother
You welcomed me, You fed me.
You embraced and protected me.
You struck down my enemies.
Like all those before me.
You guided me.
In return, I cared for you
I kept you young
When the time came.
I sadly prepared you
With deep sorrow
I laid you to rest
I bid you farewell
Sleep sweet Mother

The Power of Might

by Harry Cooke

At the age of 16-1/2, I patriotically joined the USN by lying about my age. I enlisted in the regular USN on September 25, 1943.

My first duty was eight weeks boot camp at Great Lakes. After that I was assigned to three months KP duty in Norfolk, Virginia. Then I was assigned to the USS *Missouri* in Newport, Rhode Island for two months training. One week of that training was aboard the USS *Wyoming* in the Chesapeake Bay. It was my first shipboard experience. On June 10, 1944 I boarded a real warship in the Brooklyn, New York Navy Yard. It was the USS *Missouri* and it became my home sweet home for 32 months. I was assigned to the 4th Division and was awed at the commissioning June 11, 1944 by Vice-President Harry S. Truman.

My next order was taking the *Missouri* on a shakedown cruise to Cuba. It was exciting sailing for a novice and my first venture outside America. The power and thrill of firing the guns, was devastating. The concussion from firing the 16-inch guns was breaking the glass on our signal lamps. That had to be corrected with a few other problems when we got in port.

Next we headed for the west coast through the Panama Canal. We were 108 feet wide and the Canal was 110 feet wide, so we did a lot of scraping the wall and wiggling our way through. For me it was an exciting experience.

We then headed for San Francisco and anchored within sight of the Golden Gate Bridge and Alcatraz. There were no televisions back then, so all I ever saw was a picture once in a while, so actually seeing them was to me seeing American History.

The Mighty Mo went to Pearl Harbor for Christmas of 1944. Here again I only read about Pearl Harbor in history books and the thrill of it was breathtaking.

After Pearl Harbor we went to Iwo Jima were we saw our first action. The Marines under my gun said my gun (five inch 38 gun) got the first "Japanese" plane, so I felt real good; of course, there were nine of us on my gun, so I did not do it singlehandly.

We then went to Okinawa and I kept the portside safe with my 5-inch 38 gun, but the starboard side let a kamikaze sneak through and hit

the *Mighty Mo*. On the portside the hit did not jar the ship as much as some near misses with bombs, so we did not think much of it.

Next we became the flag ship of the 38th Fleet with Admiral Halsey aboard. The battleship bombarded the coast of Japan, about 60 miles from Tokyo, for 11 days. We were on our way to bombarding the 12th day when Admiral Halsey announced the dropping of the A-bomb, and the surrender of the Japanese. I could see the Admiral when he made the announcement, because my gun was next to the bridge, and that was the first time I had ever seen Admiral Halsey smile. I also had a personal encounter with Admiral Halsey. While we were at sea, tiding up the ship, I was using emery paper to clean his hatch. He was standing about 20 feet away, looking out the hatch at the rolling sea, when my emery blew out of my hands and landed at his feet. I was scared; in those days, he was "God" with a Marine on each side of him with loaded rifles. I started to sneak away when he hollered, "Come on and get it sailor." I walked right up to his feet, picked it up, thanked him and walked out. Each step was more frightening then the action I had been in.

After that, we headed to Tokyo Bay for the surrender. It was scary, because we had a Japanese to guide us through the mines in the Bay and we didn't trust him. He turned out to be trustworthy and we got into the Bay for the surrender. We were on the portside and the surrender was on the starboard side, so I was disappointed. MacArthur came aboard the *Missouri* on the portside and it was chilling to see another "God." During the surrender I had to stay with the 4th Division on the portside, but they announced every word of it on the PX system, so I knew what was going on. One time I got excused to go to the "head," so I sneaked over to the starboard side for a glimpse of the surrender.

The finale was breathtaking with 480 American planes flying over the *Mighty Mo* wing tip-to-wing tip to show the Japanese a power of might. The 45,000 ton ship just vibrated from the roar of the low flying planes. It was beautiful!

Class Reunion

by Herbert Fahr Jr.

In early June 1954 it was time for a midshipman training cruise, the *Missouri's* last real operation before beginning a long farewell cruise to Puget Sound. All told, Task Group 40.1 was made up of 16 warships that had total of some 3,000 midshipmen on board. The *Missouri* was flagship, as she had been for the previous year's cruise. Rear Admiral Ruthven Libby, who relieved Admiral Woolridge as commander Battleship Cruiser Force Atlantic Fleet, was in overall command. Second in seniority was Rear Admiral Arleigh Burke, Commander Cruiser Division Six, who was embarked in the USS *Macon*.

Before the *Missouri* headed for Europe, she and her three sister ships: USS *Iowa*, USS *New Jersey* and USS *Wisconsin*, rendezvoused briefly off Norfolk on June 7, 1954 and did a few maneuvers. It was the only time all four ships of the Iowa class steamed together in formation. It hadn't been possible previously because of their combat assignments in World War II and Korea, nor would it be possible much longer because the *Missouri* was soon to head for the Pacific. Fireman Herb Fahr stood topside on the *Missouri* and took in the sight of the huge dreadnoughts. Up above, a photographer in a helicopter was capturing pictures of the formation steaming. In the years to come, whenever he saw one of the photos, Fahr remembered with pride that he was present for the occasion.

History Of The Mighty Mo

The Missourian Volume 2 Number 11

July 7, 1945

History is often prosaic, but not the history of the US Battleship 63, the *Missouri*, the third ship of the USN to bear the name. The first was launched January 7, 1841, mounted with two 10-inch and eight 8-inch guns. The second was launched December 28, 1901, mounted with ten 12-inch and sixteen 6-inch guns. She made a world cruise and visited Japan in 1908. The third *Missouri*, ours, was the 68th ship to be launched at the Brooklyn Navy Yard, when, on January 29, 1944, our President, then Senator Truman, made the speech of the day and his daughter, Miss Mary Margaret Truman, christened the vessel with Miss Jane Tunstall Lingo acting as Maid of Honor.

The construction of our ship was authorized by act of Congress May 25, 1939, and the order was placed on June 2 of the same year. The keel was laid on January 6, 1941, and part of the hull and firerooms had already been built when the Japanese struck Pearl Harbor on December 7, 1941. The work on the *Missouri* ceased and was not resumed until January 25, 1943 when again 5,000 yard men went to work on her daily. Injuries among men working upon her then were quite common. About nine out of every ten men had to have first aid for burns or broken bones, and more men were accidentally killed in her building than have died upon her from all causes, since she went into commission on June 11, 1944 with Captain William M. Callaghan, USN in command. On May 14, Captain S.S. Murray, USN replaced his Annapolis classmate as our captain.

The news has been released that the *Missouri* almost went to Tokyo in February 1945. She has dropped her calling cards with several Japanese clients, but strictly on business. She has not stayed for tea. Her predecessor of the same name went to Japan to cement friendship. What happened to her later? She was one of the ships sacrificed and scrapped in 1924, according to agreements made in 1919 at the Washington Conference on Limitation of naval armaments.

Our *Missouri* bites and kicks like the famous *Missouri Mule Maud* used to do. So she will live on to hasten victory, and long afterwards to assure the peace by guarding us against the piracy of other nations.

Halsey's Turkey

by Thomas F. Fluck

After months of air raids, air strikes, bombarding the Japanese, the Iwo and Okinawa invasions, Spam and sea rations, we became hardened, tough and bold shipmates. One day, preparing to relieve the watch, I passed the hatchway leading into the superstructure where the ship's dumbwaiter was, aft of the signal bridge. I spotted a large platter holding a steaming turkey, destined for the captain's quarters that Admiral Halsey occupied. The dumbwaiter inadvertently stopped at the signal bridge level. Losing composure I grabbed the hot bird, shoving it under my skivvy shirt. With greasy, burning chest I ran to the signalman's quarters, one level below. Those present, with eyes popping and dripping mouths wide open, proceeded to act like savages as they began devouring the meat. Our lookout warned of an approaching officer.

I quickly tossed the turkey over our locker where it became wedged. With a frown the officer peered in, disbelieving his sense of smell. We all appeared to be preoccupied as the lookout saluted, saying, "Afternoon, Sir!" Shaking his head the officer continued on his journey. With little hesitation three of my buddies hoisted me above the lockers, dangling me by the legs as I released the bird from its trap. Meanwhile Admiral Halsey called down to officer's mess giving what for to his personal cook. We scrapped the turkey's bones clean and carefully got rid of the evidence later. Thirty-nine years later at *Missouri's* 40th anniversary, I met Halsey's cook. Touring the USS *Laffey* (one of our Destroyer escorts in the Pacific) at Patriot's Point, South Carolina I thought the retired cook was a *Laffey* shipmate. During our conversation, Halsey's turkey came up. The cook loudly said, "Was my @#@ ever in a sling and you were the @#@ that stole it!" I then added, "Gonna put me on report?" We laughed heartily!

An Operation At Sea

by Michael Gregory

One day in 1946, a gunner's mate fell from a turret and was brought to Sick Bay, where the two doctors on board determined an emergency operation was required. It was decided that the junior officer would be charge. I don't remember his name except that he wasn't much older that the rest of us and he was a Lt.(jg).

Most of the corpsmen on board were given a duty to assist. We needed a good supply of battle lamps for the operating room and it was somebody's job to get them. Someone else had to prep the head with a shave.

The operation started with lots of Betadine (iodine) and a crayon mark to outline where the incision was to be made. The scalpel was drawn in a semi-circle and the scalp pulled down over the patient's forehead. At this point, the doctor used a few instruments to relieve the pressure that was being exerted on the brain as a result of the injury. There were no miniature TV cameras to point the way and not only that, but the operating room was slowly rolling.

The operation was a success and the patient survived. I can only guess how thankful the doctor was but there was no doubt that everyone who knew him and was there that day were extremely proud.

Once A Day Is Enough

by Al Kelley

I am probably the only 2nd lieutenant of Marines to have had a private conversation with a very famous admiral of the USN during World War II.

The "magic" of the *Mighty Mo* began this way for me and continued during my time aboard her. I arrived on the *Mo* prior to the commissioning of the ship. My first morning aboard there were no watches set yet, so I decided to take an exploratory walk. I headed aft on the portside. I spotted Admiral "Bull" Halsey approaching with several staff officers. I stopped, stepped in-board to give him ample "gangway," snapped to attention and bellowed "Sir, Good morning, Sir." I of course rendered a USMC sea-going salute. The "Bull" looked at me sort of quizzically wondering who is that noisy young Marine, walked on forward, gave me a "Swabbie Salute" ie not too snappy and very short.

I continued aft thinking if I ever have grandchildren have I got a story for them! I rounded turret three and headed forward on the starboard side. Here comes Admiral Halsey and his people. I repeated the salutation and salute. Halsey kept walking. As he passed me he said quite loudly, "Son, one of those once a day is enough!"

Being spoken to by his majesty Admiral "Bull" Halsey is still as fresh for me as though it happened yesterday and not 53 years ago. Needless to say, even though I got chewed out a little I was one happy Marine.

Mo Stories

by Warren S. Lee

In October of 1950, a newsreel camera team came aboard to photograph us bombarding North Korean shore targets. Pictures were taken of gun powder and 16" projectiles being hoisted up from the magazines and handling rooms to the turrets being trained on target. The main battery plotting room was then photographed with me pulling the trigger to fire the guns. These pictures were then shown in newsreels at movie theaters across America. In 1951 this film was used in the movie, *Retreat Hell*, starring Frank Lovejoy and Richard Carlson.

****

During our first midshipmen training course of 1949, we left Portsmouth, England and headed to Guantanamo Bay, Cuba. En route we stopped close to the Azores and had swim call. Big cargo nets were strung over the side and the crew swam off the portside aft and the officers swam forward. Marines in 30' liberty boats patrolled about 200 feet from the ship on shark lookout. Everyone was kept between the lookouts and the ship. We had an admiral aboard who also took the swim. He laid on his back and floated not paying attention to where he was. I guess the Marines were reluctant to blow the whistle. Next thing the current had him a couple of hundred yards from the ship. Then we hear over the squawk box all will witness a helicopter rescue at sea. It was quite a laugh. Had it been planned, I'm sure it would have been in the "plan of the day."

War, Boys, And Battleships

by Russell H. Lester Jr.

Although I was not in favor of the Gulf War, I guess all of us have a little war in us, or at least some story that's about or motivated by war.

One of the vivid contrasts of the old war and this new war was the smart bomb through the hotel vent, and the *Missouri* and *Wisconsin* lobbing 16" shells through their ancient guns at the beach. The *Missouri* was a very special ship to me. My father was a member of the charter crew of the *Missouri*, a "plankowner." He went through the whole war with the ship and ended up watching the surrender signed on the deck of the ship in Tokyo Bay while he was hanging from a radar mast high above. A periodic highlight of my days while he was away was getting a letter from the ship or looking at a clipping my mother had collected from some newspaper or magazine.

These clippings she judiciously collected and added to a big scrapbook that chronicled the actions and travels of the ship. The book contains Japanese money and many pictures of the ship, in one of which he claims, if you look real hard, you can see him hanging from the radar mast at the surrender signing. There is also a small replica of the Japanese flag on a card with all the signatures of those who signed the surrender. This was supposedly given as certification that one had witnessed the surrender. I could never find anything on the card that said that, but I'll take his word for it.

In those early days, with a young mind unclouded by education and experience, there was never any question about who was right or wrong. We were the good guys. They were the bad guys. My dad and his ship were on our side, and the ship was a symbol of everything that was good, right, and true in the world.

It has always stayed that way with me in spite of a growing distaste for man's warlike ways.

One of the great days of my life was a visit to the ship when my dad came home at the end of the war. Families of the crew got a special day for touring the ship, and though I knew little of what I was seeing, I was awestruck. To this day, I can mentally walk by my father's bunk, his work station, and remember the meal they fed us that day that was "just the same as the men got during their cruises." Dad laughed a little at that claim.

The ship always was there somewhere back in my mind and I didn't give it much thought until the mid-70s when I started making periodic trips to the Tacoma and Portland area. I learned the ship was in mothballs at a dock in Bremerton, and I vowed to visit it on one of my trips. As these things usually happen, I was down to my last trip in that direction and had not yet made my promised visit. I told an associate that come hell or high water, we were going in the morning. He lived in the area and had always wanted to go and also looked forward to the trip. Well hell didn't come, but high water almost did. When we started out, it was raining miserably. As luck would have it, we got there early, an hour before they opened the ship. A few minutes before they opened, I saw a naval man look out the side of the ship and saw us waiting at the

bow in the rain to get in. He was about half the length of the ship away and had to run down the pier to let us in. He looked like some type of junior officer just out of the academy doing some required duty that he wasn't particularly overjoyed with.

We ran down the dock through the rain to get to the ship and gain some shelter. I walked to the front of the ship, between the big guns, and was suddenly somewhere in the Pacific. I could feel the rumble as it crashed through the water, the guns exploding and the cold Portland rain like spray from the sea.

I went through the wardroom that was set up like a small museum, but that was all one was allowed to see.

As I was leaving the ship, I overheard the young officer turn to his companion and say, "Why would anybody run through the rain on a day like this to see this thing?"

I turned to say something to him, but realized he might never know.

In the mid-80s, when President Reagan reactivated the ship, my interests were reawakened. I wrote some letters and got my father an invitation to the recommission of the ship.

My brother, sister and I got together and gave our folks a trip to the recommission as a 50th wedding anniversary gift.

His trip to the ceremony started something I hadn't anticipated. My dad met some men who had been on the ship the same time he was, one who only lived a few miles away from his home in New York, and he became very active in the *Missouri* Alumni. I don't think he has missed a reunion since.

In September of 1988, the annual reunion was held in Long Beach, California and was scheduled to coincide with a visit to Long Beach by the ship. The crew had the Alumni on the ship

Welcome Aboard!

USS Missouri's Seal

The *trident* symbolizes sea power and it is colored gold, which symbolizes valor. The (blue and gold) *torse*, or cloth band, stands for truth and valor.

The *peace scroll*, which is white, represents the Instrument of Surrender signed by Japan on board Missouri. Superimposed on the trident are *two quills*, a red quill symbolizing the Japanese war government and a white one symbolizing the United States and her World War II allies. The (green) *wreath of olive leaves* symbolizes peace.

The *fireball* represents combat abilities. The *eight stars* represent the eight battle stars Missouri received during World War II and the Korean conflict. The *wavy* (blue) *bars* represent the sea and the Missouri river.

The *arrow shape* is the symbol for combat ability. The *ship's motto* "Vis Ad Libertatum" is Latin for "Strength for Freedom".

A Tribute to USS *Missouri*

by Maynard E. Loy Sr., GM3/c, 4th Div.

On March thirty-first nineteen hundred and
ninety two;

Marked a sad day for shipmates like me and you.

They retired the "Mo" and put her to rest,

And she will go down in history as one of the
best.

Commissioned in New York the year nineteen
hundred and forty-four,

And soon she would encounter the second World
War.

Her guns were ablaze her crew how they fought,

And the signing of the surrender on her decks
they soon brought.

Back to the States she came and with much glory,

With parades in her honor to unfold a great story.

In nineteen hundred and fifty and of need once
more,

And off she sailed to the Korean War.

How mighty her guns as she pounded the shore,

The *Missouri* was back the enemy could take no
more.

Three Battle Stars she won and her support of
the fleet,

History was in the making and hard to beat.

When it was over and the truce was signed,

She headed for home with her mission behind.

To Bremerton she would go and in mothballs
she would lay,

And three decades later recommissioned in May.

How magnificent it was the old dreadnought was
back,

refurbished to fit this new Navy's attack.

She passed her sea trials and shakedown too,

With hard work and training of a dedicated crew.

Good will cruises was her plan of the day,

But the Persian Gulf got in her way.

Saddam Hussein soon knew she was there,

With her 16 guns and missiles to spare.

Striking her targets deep inside of Iraq,

Thus supporting our troops before the attack.

Desert Storm was short-lived but even though,

It felt the force of the Mighty Mo.

She waited thirty one years to put on this show,

And now they say its time for her to go,

With her decommissioning behind and to
Bremerton once more,

Who knows her destiny or what's in store.

But I leave this with you and with much con-
cern,

That someday she may return.

So we say good-bye *Missouri* may the rest be
for you,

This battleship will surely be missed upon the
ocean blue.

Farewell to you my lady and anchors away...

for a party one night and the Alumni had a party
for the crew on shore the next night. A good time
was obviously had by all.

On the plane home, my mother wanted to
sit by the window, so my dad ended up in the
middle next to a gentleman who was on a business trip to Zurich connecting through New York.
My dad, who is one of those people who will
talk to anyone, was soon telling the man about
their trip and his association with the *Missouri*.

The man looked at my father and asked him
if he been on the ship during the war. My dad
said yes. He then asked him if he had been on
the ship at Okinawa when it got hit by a Japanese kamikaze suicide plane. My dad said yes.
"I was on the dock laying on a stretcher. I was
wounded and waiting to be evacuated. I saw the
kamikaze hit your ship."

We've probably seen the last of the *Missouri* unless they make it a museum which I hope
they'll do. I don't know how many seats are on
a 747, or why my mom wanted a window. I don't
know why Reagan decided to recommission the
Missouri or why I thought it would be a good
idea for my dad to go to the recommissioning
and I don't know why the gentleman decided to
pick that particular day, time or flight to go to
Zurich.

I don't believe in predestination and I don't
believe in magic, but I know I will never tell this
story without the hair on my arms standing up.

As a result of our renewed contact with the
ship and my deep personal interest, I decided
that I would accompany my folks to some future reunion so I could get one last look at it
before it possibly went to the scrap heap. I asked
my mom what she heard about future reunions
to be held at the ship and she assured me that
there would be many in the future. I received a
panic call from her a few days later informing
me that she was wrong and that our last chance
to see the ship would be at Long Beach, when
the ship returned from Pearl Harbor on its way
to be decommissioned.

It was not a particularly convenient time
from a business and financial standpoint to travel
from Tampa, Florida to Long Beach, but all cau-
tion was thrown to the wind and we made our
reservations. We arrived in Long Beach and had
a fine time meeting many of my dad's old shipmates.

Now to enjoy and fully appreciate the rest
of this story I am going to have to give you a
little background detail. When my dad returned
from the war in late 1945, I was nine years old.
When ever the family got together, which included an uncle who captained a tug boat in the
Pacific and a cousin who was an aircraft mechanic in the Middle East, the stories were wild
and woolly and I would listen engrossed for
hours. No matter how many times I heard the
same story, I never got bored.

One of my father's favorites, concerned a
modification he made to his battle station. He
manned an emergency diesel generator in the
forward part of the ship, when the crew was at
battle stations. Evidently there was no place to
put papers or reading material at this position.
It seems that one of the officers, my dad claimed
it was the captain, was throwing out some type
of metal desk when my dad spied it. He asked
the officer if he could have it and the officer

said sure, although he couldn't imagine what
he wanted it for. Dad took the desk and in-
stalled it at his battle station. I don't know
how that fit into regulations, but aside from
painting it red, which he was directed to
change, the desk evidently stayed in position
throughout the war.

The festivities in Long Beach included a
tour of the ship for all the men. My dad and I
were assigned two young Navy men who guided
us through the ship. I saw everything I had
dreamed about for years and the tour still ranks
as one of the high points of my life. Later in the
day the ladies were invited for tea with the officers. Evidently equal rights were not yet a consideration.

Because there were a limited number of
buses, our bus was extremely crowded on our
return trip to the hotel. I was hanging on a strap
in the back of the bus when I overheard an older
gentlemen mention that when he got out of the
service he went to Columbia University. Since
Columbia is in New York, only a few miles from
our hometown, Huntington, New York, I introduced myself and asked him where he was living now. He lived in Greenlawn, which was only
a few miles from our family home in Huntington. It turned out he had been on the ship at the
same time my dad was. I told him my dad was
in the front of the bus and that he had to meet
him when we got off.

Something I noticed during our stay was
the way old shipmates would greet one another.
They would ask rank and battle stations, then
know how close they had been to one another
when they were on the ship.

When I introduced them and they went
through this little ritual their faces lit up. "Forward Diesel" the man said, "Forward diesel" my
dad said, "the red desk" the man said, "I put that
desk in" dad said. These two guys had sat at the
same station on different shifts throughout the
war and 40 years later ended up living five miles
apart.

Like they say, you better behave yourself,
it's one small world out there.

We Are The Last

by Harold J. Loeffler

We are the last. After we are gone there will
be no more. No one will follow in our wake. For
over 100 years we were the pride of the Navy.
We were battleship sailors.

We were with Dewey at Manilla. We died
on the Maine in Havana. We manned the
dreadnaughts and sailed around the world in
Teddy Roosevelt's Great White Fleet. We patrolled the Atlantic during the "War to end all
wars" as the 6th Battle Squadron in the British
Grand Fleet.

As the battleships grew larger we grew with
them becoming more technically skilled as im-
provements in armament, engineering, and communications advanced. As technology progressed, we progressed to be the finest sailors in
the world - Battleship Sailors.

Then came Pearl Harbor. We gallantly
fought off the sneak attack. We saved what we
could to fight another day, but our losses were
devastating. Every battleship sustained damage.

The *Arizona* became a watery mausoleum for her ghostly heroic crew.

Fittingly, when the end came, the surrender instruments were signed aboard a battleship.

Extended life was given battleships with Korea, Vietnam, the Mid-East and Desert Storm and now they are needed no more. The last were removed from the Naval Registry on January 16, 1996.

We are the last. After we are gone there will no more. No one will follow in our wake. For over 100 years we were the pride of the Navy. We were Battleship Sailors.

We Remember

by Maynard E. Loy Sr., GM3/c, 4th Div.

On September 2, 1945, hence it came to be,
The signing of the formal surrender of Japanese Forces
To the Allied Powers which brought "V" for victory.
This historical event took place aboard the USS Missouri;
BB-63 and Tokyo Bay was the spot;
As our planes blackened the skies to be sure the Japanese hadn't forgot.
The crew manned the rail for the ceremony which was in store,
As the document would be signed and end the second world war.
The proceedings were opened as General Douglas MacArthur made a most moving speech,
Which would be noted with hope that pain and suffering would be put out of reach.
The Japanese came forward but with some reluctance to sign,
But Admiral Bull Halsey made sure they did not miss a line.
General MacArthur, Admiral Nimitz and other dignitaries then followed,
And with their signatures to bring to a close,
The signing of the Instrument of Surrender of the Japanese forces
To the Allied Powers and including Tokyo Rose.
This event took place 50 years ago on September 2,
A day in history the United States will always remember.

World's Greatest Warship

by Bill Egan, Journalist, USN

The year 1944 saw the aggregation of American workers add many ships of destroyer size or larger to the formidable and growing USN. USS *Missouri* (BB-63) was added to this list when she was christened by Margaret Truman, daughter of the then junior senator from Missouri, Harry S. Truman.

The *New York Times* heralded the arrival of America's newest battleship with the headline, "World's Greatest Warship Is Launched In Brooklyn." The date was January 29, 1944.

Fourth of the Navy's biggest battleships of the Iowa class, *Missouri* was destined to assume an enduring place in the history of the United States.

Seven months after commissioning, *Missouri* received her baptism of battle. On the night of February 19, 1945, *Missouri*, operating in the Iwo Jima, Okinawa and Tokyo offensive as part

of the famed Task Force 59, shot down a radar detected enemy aircraft. To her crew the ship became the *"Mighty Mo."*

Four days later another suicide plane crashed on the starboard quarter, exploding violently and throwing debris aboard main deck areas. Only superficial damage was incurred and the Japanese pilot was the only fatality.

Admiral William F. "Bull" Halsey, Commander of the Third Fleet, moved his flag aboard in May 1945, assuring the ship her share of historical fame. As flagship for Admiral Halsey, *Missouri*, at anchor in Tokyo Bay, was the scene of the signing of the Japanese instrument of surrender on September 2, 1945. This brought to a close the hostilities of World War II.

Following the close of the war, the *"Mighty Mo"* remained the only US battleship on active duty, as one by one, her sister dreadnoughts joined the mothball fleet.

For five years *Missouri* operated with reduced crews on special missions to Turkey and Brazil and on numerous midshipmen and reserve training cruises. One such mission was to return the body of the deceased Turkish Ambassador to Istanbul; another carried President Truman and his family home from a special hemispheric conference in Rio de Janeiro. In effect, the nation's most historic battleship became a floating "White House."

Two months after the outbreak of hostilities in Korea, on August 13, 1950, *Missouri* interrupted her midshipmen cruise and sailed for Korean waters where she joined Task Force 77.

In December 1950, United Nations armies walked into one of the biggest ambushes in history and began running for their lives. On December 23, exhausted Marine Corps leathernecks, carrying their sick and wounded, stumbled onto the beach at Hungnam and found *"Mighty Mo"* and a force of cruisers and destroyers waiting to enfold them in protecting arms of fire. *Missouri's* guns roared a curtain of steel around the beachhead through which the enemy could not penetrate.

In the weeks that followed, *Missouri* cruised unchallenged up and down the coastline, demolishing bridges, trains, tanks, and troops. Generals began calling her "the best infantry weapon the Army ever had."

During the time *Missouri* spent in Korean waters, she steamed more than 80,000 miles and fired 7,300 tons of ammunition at North Korean installations.

Missouri was decommissioned and carefully preserved in February 1955 to rest at the Puget Sound Naval Shipyard in Bremerton, Washington for three decades.

During her inactive years, *Missouri* continued to serve the nation - some 180,000 visitors toured the battleship to view the surrender deck each year.

In May 1984, *Missouri* was ordered to once again join the Navy's active fleet. The battleship was delivered to Long Beach Naval Shipyard for a two-year program for modernization to the needs of today.

Guns of "Big Mo"

Submitted by William F. Mowder

Tokyo, Dec. 18 (AP) A fiery bombardment by warships and field artillery today held at bay another Red Chinese assault on the United Na-

tions' tiny Hungnam beachhead in northeast Korea.

(The US Battleship *Missouri*) arrived offshore. Its 16-inch guns and one-ton shells were a welcome addition to the curtain of fire shielding hard-pressed units of the US 10th Corps.

Major-General Edward M. Almond, 10th Corps commander, obviously was pleased by the intense firepower hurled at the masses of Chinese infantry pressing on Hungnam port from three sides.

"Things are going just the way we planned them," he said. "Now every time the Chinese Communists dig in, we hit them with artillery concentrations, mix them up and knock them out. That's something we have been unable to do before."

At no point on the port's defense arc had the Chinese been able to punch through. Observers said the Reds apparently hand not sent many troops into bomb and shell-shattered Hamhung, industrial city six miles northwest of Hungnam, Hamhung was abandoned to the Reds on Saturday.

An estimated 25,000 Chinese pressed against the beachhead rim. Another estimated 75,000 were moving up in the snow-mantled hills west and northwest of Hungnam.

Arrival of the battleship *Missouri* increases the range and effectiveness of naval fire. The *Mighty Mo's* guns have a range of 20 miles - far enough to reach the white hills sheltering Chinese rear positions.

The *Missouri* entered the Korean War on September 15 with an intensive shelling of Samchok on the East Coast. This was after an 11,000 mile dash from Norfolk, Virginia. She appeared off Inchon September 21 to help cover the Allied West Coast landings and then returned to the northeast Korean coast. The 45,000 ton warship last was reported in Korean waters November 7.

General MacArthur's war summary credited combined naval and ground force bombardment with breaking up a pre-dawn assault against perimeter positions west of Hungnam - evidently an attack intended in great force.

Grandfather's Diary

Submitted by Jack Norton's grandson

The following is an entry taken out of a diary kept by my grandfather, Jack Norton, He served on the Missouri during World War II and kept a detailed diary. Although I never met my grandfather, I feel as though I've gotten to know him through his diary.

March 18, 1945 - What a morning! We shot down one Judy, two others were shot down by the planes. Our large strike went in at dawn and not back yet. Their fleet is ready to come out but we are blocking their way. The *Enterprise* took one bomb hit, *Intrepid* took a devine wind. Neither hurt too much, and no news as to how many people hurt. A dive bomber was just shot down. They could see land at one time today, this should determine the Japanese fleet.

Bye darling. God had better be with me today. Headed for sky control. 2000 back again, I don't believe I ever went through such a day. Under attack all day. I was looking at a cloud, when

Jack Norton with wife Bernice and daughter Judy.

all of a sudden, a Japanese wind came in for a kill - guns blazing! We knocked it down. It burst into flames and he drove it at a nearby carrier. The pilots bailed out and the ships shot them as they were coming down. We don't want prisoners.

Another plane came in and the five-inch took care of it, it was a radial engine Judy. *Intrepid* shot one headed for her, and it just missed the *Alaska*. One headed for the *Wisconsin* but exploded in mid-air. The *Wisconsin* shot down one of our planes, but the pilot was saved. The Japanese have two battlewagons, two or three C1. We knocked one down off our fantail, and it blew up and showered the *intrepid* with gas.

We have been at "GQ" for 15 hours so far and it's still early. The *Enterprise* and *Intrepid* had flags at half mast as they had to bury their dead. Poor devils.

One Japanese plane with a 500 pound bomb dove into the carrier *Franklin*. The planes on the flight deck with bombs and gas exploded. Eighty men jumped into the drink! Some were blown in! The carrier is worth saving, that is about all. I was standing on 05 level when I saw it get hit, I hope not too many were killed, it looks very bad though. Our strike got two Cl cruisers, 55 Zeroes, and many planes on the ground. Our position is now 100 miles south of Japan. I will soon need matches to hold my eyes open. We had K-rations today. No wonder the dogs back home are going hungry.

Memorandum
For the Members Of The Third
Fleet Naval Landing Force
Regiment

by L.T. Malone, CO

We have been chosen, largely by luck, to represent our US Navy in the occupation of Tokyo. There were close to one quarter of a mil-

lion officers and men in the Third Fleet to pick from and we got the nod. We are honored to have this opportunity to represent our Navy in this occupation. Many others will follow us in after we have squared things away, but we make the initial impression and, mark you well, it will be one of the great first impressions of history.

Before you can get into a sweat about impressions made you have to know what you are to do. It is this!

The two assault battalions now in the monitor and a nucleus crew and service battalion loaded in two APDs, will occupy and take over the Yokosuka Navy Yard. This is the Empire's 2nd largest naval base. It is the headquarters for their First Naval District and it is in inner Tokyo Bay just short of Yokohama and Tokyo proper.

We are but a part of the occupation force which takes over this Yokosuka peninsula.

The 4th Reinforced Combat Regiment of the famous 6th Division of Marines under Brigadier General Clement, USMC will spearhead the attack as a combat regiment. The British Navy is represented by a battalion of Royal Navy Tars and a battalion of Royal Marine Commandos. All of us are under the command of the Marine General. We are proud to serve under him.

We are operating in fast company - men of two nations that are seasoned in combat and inspired by many memorable and historical campaigns. We in our working uniforms are the US Fleet. The fleet that Admirals Halsey and Spruance, under the able leadership of Fleet Admiral Nimitz, have churned the Pacific waters with to keep our air, sea and sub-surface groups relentlessly battering a formidable foe to his knees. The foe has capitulated. We have tested him and we now go into his inner sanctum sanctorium to raise the American Flag and occupy his territory. Our responsibilities are great and our obligations tremendous.

In theory we are entering surrendered territory that has been evacuated by enemy armed forces. In fact that may not be wholly true. We go prepared to take over this naval base in a formal, dignified manner. We hope that no misguided fanatics will make it necessary for us to use arms. But we must keep in mind that our enemies have previously been treacherous. Their codes are different than ours. We may have a little difficulty. However, we do not anticipate trouble and we do not seek it. Our mission is to occupy the Yokosuka Naval Base. We shall carry out that mission. Each officer and man can well appreciate our responsibilities to the nation and to our brothers in arms. We are on parade with full pack. We are not a spit and polish organization but we will conduct ourselves and insist that our messmates and shipmates conduct themselves in a manner befitting this memorable occasion and the Nation and fleet we proudly represent.

Our naval regiment has 80 competent officers. Obey them explicitly. Keep quiet, act military and make the US Marines proud of you.

We Remember

by Louis S. Pisani, S1/c, FA Div.

I was assigned to the FA Division and my battlestation was at the toughest point on the ship, the conning tower, on a platform that surrounded

the main director for the 16" guns. They assigned me up there as a target designator. It was an instrument with large binoculars to help pick out enemy targets. This platform was called air defense control. The air defense officer controlled everything from up there and had a good view of all the actions. I remember watching all the planes taking off from the carrier for air strikes at the enemy targets. I saw the kamikaze in action, and I saw them knocked out of the sky and hitting our ships. I remember when an air strike was returning from a run at Japanese targets. The planes were flying back in formation and somehow, I don't know how, the Japanese suicide plane flew in undetected with this flight. He may have flew above it or below it, but our radar or IFF didn't pick him up. As our planes were landing on the USS *Franklin*, the suicide plane dove out of the clouds following our planes in and before it could be detected, it dove in just under the hangar deck of the *Franklin's* fantail and there was a tremendous explosion. I was scoping our planes as I usually did with my binoculars and recognized the suicide plane as Japanese followed it right into the ship. It was a very disheartening sight. I also followed the plane in that hit us and my heart was pumping, because when you hear the 5" going off, then the 40s taking off, then hear the 20s, you know he is close. Hit, he tumbled forward and burned. The smoke rose up by our tower; I think I was waiting for it to explode, but it didn't. If it had it would have taken a lot of lives. We had a speaker to monitor the aircraft on their air strikes. We could hear the conversations between the aircraft and to the ships. That kept our hearts pumping too.

These memories will last and at my age I am glad I can still remember, but then the "Mo's" motto is "We Remember."

The First Known Helo-Ship
Operation

by William A. Pitcher

Although the period in which I served, October 1946 to February 1948, in the USS *Missouri* (BB-63) was after World War II, it was nevertheless interesting because the ship was near full complement and fully operational.

Probably the most noteworthy event was the trip to Rio de Janero in the late fall of 1947, where the *Missouri* took part in the protocol of President Truman's role in the Inter-American Conference and then had the President, his family, and senior staff on board for the trip back to Norfolk.

However, the event most remembered by me was the Cold Weather Cruise to the Arctic Circle in November 1946, shortly after, I reported on board.

A couple of days after getting underway from Norfolk, it was late afternoon when the word came over the 1MC, "Fire in No. 1 Turret!" As newly appointed fire marshall, I headed for the scene, picking up members of the fire party on the way. Quickly reaching the lower handling room in the turret, we found the deck so hot that we could barely stand on it. This condition had prompted the turret watch to report to the bridge.

The first consideration was to remove the powder bags from the scene, which was expeditiously handled by the 1st Division.

Since there was no smoke or indication of fire, the possibility of a broken steam line in the space below the affected area was considered.

The chief engineer, who was at main control, was contracted about this possibility. He replied that the closest steamlines were the fuel oil heating lines in the adjacent fuel tank abaft the void according to his diagram. However, he quickly and correctly took a back suction on the steamline. Shortly thereafter, the temperature subsided and a report was made to the bridge and combat.

At that time, the first lieutenant and the XO burst upon the scene, having been lost, and taking about 30 minutes to reach the lower handing room.

I was able to report that everything was under control. When the area had cooled down, and we were able to go into the void, it was found that a loop of the steam line had extended into the compartment (an error by the shipbuilders), and had ruptured.

Later, I was to reflect on what would have happened to my butt and career if the first lieutenant and XO had not taken so long to find their way to the bottom of No. 1 Turret.

There were two other memorable incidents during the cruise. One occurred during a nighttime star-shell illumination exercise with a cruiser, the USS *Little Rock*. A star-shell correction failed to get into the computer, resulting in a 5" star shell hitting the *Missouri* amidships at the 01 deck, the exact geographical center of the ship. The shell hit in a nest of acetylene bottles, causing a nasty fire that damaged several staterooms and compartments, two personnel were seriously injured. Damage was considerable, but the ship-fitter gang repaired the damaged area so that when the *Mo* returned to Norfolk, there were no scars visible.

The other incident occurred while at anchor in the Bay of Argentina. I had the afternoon watch as OOD when a Coast Guard helicopter approached. The pilot notified us by radio that he intended to land on #2 turret.

A quick check found that the commanding officer had given his permission for this while at the Officers' Club the previous evening. Landing on the turret was one thing, but keeping the helo there, with the wind blowing, was another.

A team of daredevil volunteers was formed that grabbed the helo when it landed, and secured it in place.

Captain Hill then appeared and climbed on board the helo, where upon the *Intrepid* helo pilot took off and headed for the beach. This was the first known helo-ship operation. At the fleet maneuvers in the Spring of 1947, helo-ship operations for mail and personnel transfer became an integral part of operations.

Surrender Gig

by Carl Reisman

Scores of news media personnel were coming to record the forthcoming formal surrender of Japan scheduled aboard the USS *Missouri* on September 2, 1945. The painted wood decks were laboriously scraped to show off the ship to its best advantage, but there was not enough time for the sun to bleach the teak wood to its nearly white color. The order went out to repaint the decks as fast as possible so they would be dry for the ceremony, and also only to paint the starboard side of the ship since we were short on both time and paint. The news photographers were only going to be allowed to photograph the ceremony which was starboard, so it did not matter too much if the portside of the ship was not freshly painted.

The decks were painted, using all the paint aboard the *"Big Mo,"* plus additional paint needed to complete the task obtained from various other ships anchored in Tokyo Bay. The paint barely got dry to the touch late on the evening of September 1, 1945. I know, because I checked the paint on the quarter deck which was maintained by the 8th Division just before I went to my cabin to turn in and get some sleep. Before I was able to get undressed, I was ordered to report to Admiral W.F. Halsey who was in a meeting with Captain SS Murray in the Captain's quarters.

Upon my arrival, a very busy Admiral Halsey said, "Reisman, the Captain tells me that you would be a good man to send to get the Japanese surrender party and bring them to the *Missouri*. Have them alongside ready to board at 0855 sharp." I replied, "Aye, Aye, Sir" and I left. Outside the Captain's quarters, I realized that I did not know where the surrender party was or where I could meet them, and I certainly was not going to go back and ask Admiral Halsey.

I finally obtained that information from the Communication Information Center (CIC). The Japanese were being picked up by one of our destroyers that was to anchor in Tokyo Bay in the early a.m. I obviously needed a small vessel, but when I requested one from the officer of the deck, I was told that he could not provide one for me. He was having his own problems getting enough boats to get all the people aboard for the ceremony and I would have to make my own arrangements. I requested permission to send a message to another ship by blinker and was told by the Officer of the Deck that I could do that. He wished me good luck in trying to find a boat and then added that if I had difficulty getting one, to please leave him alone because he had his hands full and was too busy to help me.

In my tour of active duty during World War II, I commissioned three ships, one of which was the USS *Iowa* which was anchored 1,000 yards off of our starboard bow. I served aboard the *Iowa* before I commissioned the USS *Missouri* which is a sister ship of the *Iowa*. I had the signalman on the bridge send a message to the *Iowa* which read: "Request permission to borrow the captain's gig to pick up the Japanese surrender party and bring them to the *Missouri*" ...and I gave my name and rank. The message was acknowledged and I was told to "stand by." Then I received the tongue in cheek answer which read: "Permission granted to borrow the Captain's gig to pick up the Japanese surrender party even though you did desert the *Iowa* for the *Missouri*."

The Captain's gig from the *Iowa* arrived at the *Missouri* at 0800 sharp which allowed 55 minutes to go and pick up the Japanese and make the return trip back to the *Missouri*. When we arrived at the destroyer, a rope ladder was already hanging over the portside of her forward main deck and the Japanese were waiting to climb down to board the gig. The choppy water caused the boat to bounce against the side of the destroyer. The first Japanese person to climb down the rope ladder spoke English and was the interpreter in addition to his other official duties. The Japanese party consisted of six military officers and three civilians including Mr. Mamoru Shigemitsu, the foreign minister. He was dressed in formal attire, including a top hat, tails and a white vest which had yellowed with age; his whole outfit smelled of camphor moth balls. He had a prosthetic leg and lost his balance when stepping from the rope ladder onto the gig. I grabbed him and pulled him into the boat. After all, he had not signed the surrender document yet.

Returning to the *Missouri*, I recalled that I had been told to be alongside the gangway ladder at 0855 sharp. I had allowed enough time for the journey and a bit too much, so we circled the *Missouri* even though she had not been painted on her portside. The Japanese had a chance to see what the rest of the world and news media did not, the portside which had not been repainted, but the extra time permitted us to arrive exactly at the gangway of the *Missouri* at 0855.

The surrender was on the *Missouri*, but all ships were ready and willing to help in any way possible to bring the war to an end, whether in fighting the enemy, sending over needed paint or lending an extra needed boat. After all, we were all in the war together.

Chronology of Events

by Carl Reisman

Seven days before the start of the surrender ceremonies, the USS *Missouri* took aboard Japanese emissaries and pilots to obtain critical information on minefields and harbor conditions in Sagami Wan and Tokyo Bay, where the United States were to enter.

Carrying Admiral Halsey, the flagship of the Third Fleet, sailed into Sagami Wan on August 27 for a rendezvous with a Japanese destroyer carrying Nipponese naval officers and pilots.

The USS *Nicholas*, a destroyer, which moved ahead of the flagship, took aboard 18 Japanese by small boat transfer. Over bitter protests, the Japanese were relieved of their beloved Samurai swords and daggers. The *Nicholas* then moved alongside the *Missouri* and transferred the Japanese by boatswain's chairs.

Peering from the bridge, Admiral Halsey grinned as he watched the Japanese arrival aboard ship. Charts of Sagami and Tokyo Wan, as well as other Japanese waters, were scrutinized to specify the location of the minefields. The information proved accurate. The fleet was able to move into Tokyo Bay for the surrender ceremonies without incident.

On the morning of September 2, 1945, Admiral of the Fleet Chester W. Nimitz, USN, arrived on board at 0802 with his staff. High ranking Army and Navy officers began coming

AV Division.

aboard shortly after 0800. Generals Stillwell, Krueger, Hodges, Spaatz, Kenney, Doolittle and Eichelberger were among the Army leaders present. General Jonathan M. Wainright was also present.

The Navy was represented by Admirals Halsey, Turner, Towers, McCain, Lockwood, Sherman and others. Lieutenant General Geiger represented the United States Marine Corps.

At 0830 representatives of the Allied Powers began coming aboard. The Republic of China was represented by General Hsu Yung-Chang; the United Kingdom by Admiral Sir Bruce Fraser, GCB, KBE; the Union of Soviet Socialist Republics by Lieutenant General Kuzma Nikola evish Derevyanki; the Commonwealth of Australia by General Sir Thomas Blamey; the Dominion of Canada by Colonel L. Moore Cosgrave; the Republic of France by General Jacques LeClerc; the Commonwealth of New Zealand by Air Vice Marshal Isitt; the United Kingdom of the Netherlands by Admiral Helfrich. Scores of photographers swarmed the ship.

A few minutes before 0900 General Douglas MacArthur, Supreme Commander for the Allied Powers, arrived aboard. He was met by Admiral Nimitz, Admiral Halsey and Captain Murray.

The Japanese party consisted of Foreign Minister Shigemitsu, General Yoshijiro Umezo, Chief of Staff, Japanese Army Headquarters; Katsuo Okazaki, Director General, Central Liaison Office; Shunichi Kase, Director Number One Government Information Bureau; Lieutenant General Shuichi Miyakazi, representing Army General Headquarters; Major General Yatsuji Nagai, Army Staff; Rear Admiral Tadatoski Tomioko, representing Naval General Headquarters; Rear Admiral Ichiro Yokovama, Navy Headquarters, and Captain Katsuo Shiba, Navy Headquarters.

The Japanese delegation mounted the gallery and took positions facing aft towards the assembly of representatives of the Allied Powers.

Promptly at 0900, General Douglas MacArthur came from Admiral Halsey's cabin on the *Missouri* and took his position in front of a battery of microphones, through which this drama was to be heard in Washington, London, Paris, Berlin, Rome, and in all the cities, towns, hamlets and farmhouses of the world where radio could be heard.

General MacArthur spoke briefly, unemotionally, describing the importance of the event in its bearing on an end to a war that sprawled over the vast reaches of the Pacific. He wasted little time, as no general or admiral does, on bombastic theory. He described the military situation and let it go at that.

At the conclusion of his speech, General MacArthur requested the representatives of Japan to advance and sign the documents which were spread on a green, cloth-covered mess table.

Mr. Mamoru Shigemitsu, Foreign Minister of the Japanese Government was the first to sign the instrument of surrender for Japan. The world sat with its ear cupped intently to the radio. This was the first time the average citizen had been able to sit in on the details of a war.

General MacArthur then announced he would sign for the Allied Powers collectively after which each representative would sign for his country. The signing proceeded in an orderly fashion. When all had signed, General MacArthur announced that it was his purpose to see to it that the terms of surrender were carried out. Then he announced: "The ceremonies are completed!"

The success of the efforts of the Missouri's personnel was recognized in the congratulatory message of Admiral Nimitz who praised the efficiency whith which the entire day's proceedings were carried off. To all hands went a "Well Done."

Life on the Mighty Mo

by H. Kirby Riner

At 31 years of age I had a good life, a beautiful wife, two lovely daughters and a good job as foreman for a local subsidiary of Martin Marietta. The Second World War was escalating, and I was frozen to my job, but my thoughts were on my friends, relatives, and former schoolmates serving in the armed forces. Some had already given their lives. I felt a need to contribute to my country so I advised my employer I would be quitting to join my fellow countrymen. The company released me the following day and I left Martinsburg, West Virginia with the next draft contingent, enlisting in the USN.

After training in Great Lakes I went to Norfolk, Virginia for SOSU. I stood muster, had guard duty, swept up hangars, and fueled and beached our Kingfisher planes, even going up some both day and night. But I yearned to do more. A week of this went by and while standing muster one morning our officer asked for volunteers for duty on a battleship or one of two cruisers. I got the battleship, not knowing it was the USS *Missouri* until I later received my orders, and I was aboard the *Mighty Mo.* I slept in a hammock that night full of apprehension and excitement, never dreaming that I would see history in the making.

The next day the ship's gear was assigned to me, and I was assigned to the AV Division quarters and personnel. I was then assigned to the AV starboard catapult where I would be manning the phones, and this would be my battle station.

We pulled away from port and headed out to sea to join our fleet in the Carolines. I remember seeing the new carrier, the *Randolph*, off our portside and forward of us. Later that afternoon a Japanese plane came in low from the Japanese-held island of Yap and hit the Randolph. There were a lot of explosions and damage; and with it came the realization that we were in the war zone.

The first plane shot down on a fleet operation was picked up and shot down by our ship late at night. This put the USS *Missouri* in good standing with the rest of the fleet. During our first operation at Okinawa, our big guns bombarded the Japanese strongholds and helped establish a beachhead for the US landing. The second time we were there Admiral Halsey was aboard and the fighting was in the Naha area. Our big guns again gave assistance. We met no opposition. She truly was the *Mighty Mo.*

One day a Zeke plane hit our ship on the starboard side aft, I was there manning the phones and I could not see the superstructure as it was engulfed in smoke. Damage control had everything taken care of and under control in a matter of minutes. I would like to say our damage control unit was the best. A gun from the plane lodged in the barrel of one of the aft five-inch mounts and a piece of the wing went forward. Our AV metalsmith, Ed Sadowsky, gave me a piece of the Rising Sun. The pilot's body was found under a basket that contained life floats and the next day his body was brought topside, starboard aft, and he was given a military funeral and buried at sea.

My duty during flight quarters was on the phones. After the pilot was in the plane, had put it through its paces, and checked his instruments, the gunnery officer put off the charge that shot him down the catapult, I then reported by phone that the plane was airborne. I was on the phones when the 16-inch guns were fired. I realized that I could hear a clicking noise on the phone an instant before they fired the salvo. Possibly this occurred when the breech was closed. So I would hear the click, hit the deck and all persons in sight would hit the deck. One time I did not hear the click (warning me) and I got a flash from the catapults steel track. I couldn't see so I slumped to the deck saying a prayer and gradually my sight returned.

I always stood Dawn Alert whenever it sounded and mostly was the first one out of the hatch. One particular morning it sounded, and out I went. It was quite dark and rough, and I was getting very wet. I put on my headphones over at the starboard catapult and grabbed a line out of the plane and tied myself to the railing and to the SCI Seahawk plane. I immediately called the bridge, getting no answer. I called again and again, still no answer. Daybreak came and the storm began to subside some. Finally after repeated calls, the bridge answered and said, "What the hell are you doing down there?" I replied that I was manning my station as Dawn Alert had been sounded. He answered "Damn, we secured it right after we sounded, it was too rough out there." So much for being first out of the hatch - but I said a little prayer that I had not been washed overboard. I was not relieved at 8:00 a.m. as General Quarters was sounded, so my friend shipmate Sadowsky brought me some breakfast.

One day General Quarters sounded and I was running across the deck for my phones when a Japanese plane came into view. He was about 10 feet above deck level on the starboard side and headed aft. I could see the pilot real good as he and I were going the same direction. His plane hit the AV crane and when I got my phones on I reported the damage to the

bridge. Results of the crash threw the pilot into the sea and he drowned.

I remember the day the carrier *Franklin* got hit. It was off our portside just beyond the horizon. We could see the smoke billowing up from our position. We heard how badly she was damaged and of the lives that were lost. I was concerned as a friend, Doug Vorhees, from my hometown was aboard the *Franklin*. About a month later I learned Doug had been transferred to another ship and was safe. I said a prayer of thanks.

Being topside so much I saw a lot of pilots come back to land on their carrier. Some were so damaged that they would land in the sea and be picked up rather than risk damage to their ship. Others would come back and land right on target and you had to admire them for their ability because that carrier is a mere dot in that vast ocean. I believe Navy pilots are among the very best.

There were some, however, who did not come back. One was Lieutenant Frothingham from our AV Division. One morning as he was out on a mission I saw his plane disappear on the horizon. Word came back that he suffered a severe medical emergency and had to put his plane down. His plane capsized but he was afloat and alone in the high seas. A destroyer went immediately to his aid and hooked onto him. But the waves were so high he was washed away and couldn't be recovered. Lieutenant Frothingham was one of those persons I most admired. He was a quiet, handsome man, looked like an aviator and one of the best. The entire AV Division and crew grieved this loss. My thoughts and prayers immediately went out to his family. I thanked God I had been privileged to know him.

There was a lot of camaraderie among the men on board the USS *Missouri*. Some of us have stayed in touch. Some are in ill health or deceased. But those of us who are still living will always have the bond formed on the *Mighty Mo.* I left her in December 1945. Since that time I

Navy Blue

Submitted by Denny W. Wilburn

Say Girl, I saw you sneer just then,
Didn't I look good enough for you?
I'm not one of your class you say,
I wear the Navy Blue.

You think I'm not fine enough,
For such a girl as you,
Men who wouldn't hold your hand
Have worn the Navy Blue.

You bar us from the theaters,
And from your bar rooms too,
Where there's room for everyone,
Except the Navy Blue.

How many folks in civilian life,
Will take the time to think,
That sailors do other things
Besides smoke, swear and drink?

When we are dead, when we are gone,
When life's last cruise is through,
We'll not be barred from heaven's gate
For wearing the Navy Blue.

So when you see a sailor boy,
I'd smile if I were you.
No better men are made by God,
Than the men in Navy Blue.

had an occasion to meet Captain Callahan and I told him how proud I was to have served under him on that great ship.

I have been aboard the USS *Missouri* once since I left her and that was in 1983 in Bremerton, Washington. I will always have great memories of her! My only regret is not keeping a daily log of event and life on board. I am so very proud to have served my country aboard the USS *Missouri.*

USS Missouri 7th Fleet Champions.

Skin Deep

by Bob Schwenk

After arriving in the South Pacific in March 1945, I was assigned to mess cook, and put in charge of #1 chow line. After the Iwo Jima campaign, we pulled back to Mogmog Island. Being from California and liking to swim, I would take two of my mess cooks to the fantail while at anchor. I would look around and have the mess cooks watch for the officer of the deck. When he was turned the other way, I would take off my clothes, get up on the 40mm gun tub, dive into the ocean, swim around awhile, then climb up the ladder on the stern, put on my clothes, and go back to work. Really cooled a guy off!

Then we took off for the invasion of Okinawa. A Japanese kamikaze plane came in off the starboard stern of the *Missouri*, was a little bit high and his left wing hit the aviation crane. He spun down astern of the ship and blew up.

We arrived back on good old Mogmog Island and I decided to go swim as before. I dove in, swam awhile, and when I got to the rungs of the stern ladder, at about four feet up, all the rungs were smashed against the ship by the kamikaze plane. The guys on deck are yelling at me to come up, and I am trying but can't get a good hold of the rungs. I am thinking I am going to have to swim around to the after gang way with no clothes on and ask the officer of the deck for permission to come aboard. He will ask "Where are your clothes?" And I will tell him, "The fish must have eaten them when I fell overboard. I finally jumped out of the water, grabbed one of the rungs, and clawed my way back up the deck. That was the last time I went swimming off the fantail.

When the *Missouri* was in Long Beach in dry dock, I went down to the ship, looked over the stern, and was trying to see if my fingernail and toenail scratches were still there. I guess over the years they have worn off.

Kamikaze Attacks on the Missouri

The Missourian Navy Day

October 27, 1945

A sheered-off wing lying near a 5-inch gun mount, a machine gun driven through the barrel of a 40mm gun, a charred body on the deck, hundreds of broken plane parts...."

These were the after-effects of the most determined kamikaze attack made on the *Missouri.*

The suicide pilot who drove his Zeke on the ship on April 11, 1945 had hoped to damage or cripple the ship. But beyond a small fire started by gasoline the ship went unscathed. Not an officer or man was hurt.

The plane, attacking in the early afternoon at low level, was taken under fire by the ship's guns as he made his hara-kari run on the starboard quarter. With pilot riddled and probably killed by the terrific hail of bullets from the *Missouri's* guns, the plane kept boring in until it crashed into the ship within a few feet of the AA gunners.

Not a man left his gun as the plane dove aboard. One wing sheered off and flew forward, landing inboard of 5-inch gun mount No. 3 where the gasoline from the shattered wing burst into flame. Clouds of smoke and fumes were sucked into the fire room by the main ventilation intake nearby, but the fire was quickly put out by a party led by Lt.(jg) O.D. Scarborough, Junior Officer of the Deck.

Damage Control officers quickly assessing possible hurt to the ship, found the main deck aft littered with fragments of the Zeke. A crushed remnant of the pilot's body, thrown clear of the wreckage and found lying on the deck, was given burial. The plane's machine gun was driven through the barrel of one of the ship's 40mm guns, so strong had been the force of the impact.

Other kamikazes, spurred by a desire to avenge the *Yamato* which had been sunk by planes from the Third Fleet, attacked the *Missouri*, but none other was able to penetrate the screen of fire which the ship's gunners threw up around her.

Coast To Coast In 9 Hours and 20 Minutes

The Missourian Navy Day

October 27, 1945

Coast to coast in 9 hours and 20 minutes. That's the time it took the *Missouri* to transit the Panama Canal on October 13, 1945 on her historic voyage back to the East Coast.

The Flagship began to move up the channel from Balboa at 0701 Saturday and reached sea level on the Atlantic side of the Isthmus of Panama at 1621.

Carrying Rear Admiral John R. Beardall, Commandant of the 15th Naval District, high ranking officers of the Army, and distinguished visitors from our Latin American neighbor countries, the *Missouri* made the transit with minimum difficulty.

A stanchion was knocked down on the starboard side; another was bent almost to the deck. Chips of concrete were scraped off the walls of the towering lock chambers. But the four pilots who had the ship in charge: Captains Majelton, Hearn, Redman and Saunders, snaked the *Mighty Mo* through the Big Ditch in short order.

Three tugs helped maneuver the ship into the entrance to the first flight of locks, the Miraflores at 0801. There the raising of the ship to the level of Miraflores Lake was halted briefly while the guests came aboard.

Full honors were provided as Rear Admiral and Mrs. Beardall, Lieutenant General G.H. Brett, commanding general of the Caribbean Defense Command and Panama Canal Department, 12 other generals, other Army officers, the Venezuelan and Peruvian ambassadors to Panama, and Panama Canal officials came aboard. Many of the guests were accompanied by their wives and children. Thousands of residents of the Canal Zone lined the lock to cheer the ship as she moved through.

At 0925 the ship left the locks and sailed into Miraflores Lake, reaching the Pedro Miguel locks at 1008. There most of the visitors left the ship while shouting and cheering spectators hailed the *Missouri* from the banks. Hundreds of sailors' white hats sailed ashore to be eagerly grabbed by the hundreds of small boys and girls in the throng.

As the *Missouri* left the locks at 1102, a tug took her bow line to help her negotiate the winding course of the canal through Gaillard (Culebra) Cut. While an Army officer at the loud speaker related facts of interest about the canal, the ship's company eagerly scanned the shores for the alligators and iguana lizards reported to frequent them.

All hands marveled at the sheer determination and courage required to dig and blast the passage through the saddle of rock which once connected Contractors' Hill on the left with Gold Hill on the right. At Gamboa, where the massive cranes and dredges used to keep the canal clear of steadily accumulating silt and rock slides lay at anchor, the tug cast off and the *Missouri* slid along smoothly into Gatun Lake.

Visibility vanished in a short, fierce rain squall which cleared by 1437 when the ship entered the Gatun locks, the last set. Here again a huge crowd had been patiently waiting for the ship's arrival and cheered, waved, and skylarked as she went down, down, down the 85-foot descent to sea level. Scores had bunches of bananas and threw them aboard in exchange for white hats. Girls wrapped messages around magazines and tossed them to grinning sailors on the weather decks.

At 1621 the *Missouri* finished her transit of the canal proper and entered Limon Bay at 1655. Twenty minutes later the pilots left the ship and Captain Murray took over her direction for the last leg on the "Long Voyage Home." The breakwater at the entrance to the bay was cleared at 1742 and at 1829 the Ship's Log reports the ship on a course for Hampton Roads, Virginia, with Colon fading in the distance, 21 miles astern.

The *Missouri* had a foretaste of what to expect in New York on Friday when thousands of visitors swarmed down to Pier 18 at Balboa to visit the ship.

Among the 35 special guests were: Dr. Ricardo Alfaro, Panamanian Minister of Foreign Affairs and members of his family; Walter Donnelly, Charge d'Affaires, U.S. Embassy, and his family; and 14 members of the Mexican delegation to the Inter-American Congress of Lawyers including Jose Ortiz Tirado and Hernando Hilario Medina, magistrates of the Supreme Court of Mexico.

But while these distinguished visitors were being guided through the ship by officer escorts, thousands of Americans and Panamanians rushed to board the *Missouri*. From the main deck to sky control the visitors: men, women, boys and girls, walked, ran, and scampered over the Flagship. Many hundreds more were regretfully turned away when it became necessary to close the ship to visitors.

Denied admittance, they continued to line the pier and gape in wonder at the massive battlewagon, eagerly pointing out to each other the big rifles and bristling AA guns and speculating on the purpose of the detecting and pointing equipment of the ship.

While Panamanians were visiting the ship, her officers and crew were repaying the courtesy by visiting in Panama. Shops, stores, bars, night clubs, cathedrals, churches, the parks and public buildings, all received a friendly, although hasty call from the ship's company. As a result hundreds of mothers, sisters, wives, and sweethearts will be the proud owners of alligator bags, perfume, dresses, and linen, souvenirs of the *Missouri's* one-day visit to the canal.

Rig For Church

by Chaplain Roland W. Faulk

Any portrayal of the life of the men aboard the *Missouri* would be incomplete without reference to the place accorded religion. From the day of commissioning, when the blessings of God were invoked on the ship, throughout all the days of combat and long and monotonous cruises, the men of the *Missouri* have enjoyed one of the blessings of liberty, the right to worship God.

The *Missouri* has carried, since commissioning, both Protestant and Catholic chaplains. Although many services of worship have been interrupted or postponed because of battle conditions, the services when held have always been well attended. Rarely has it been necessary to call off church services even though services were sometimes held at odd hours. During the Iwo Jima and Okinawa campaigns and the later bombardment periods, Church services have been held. Men of Jewish faith have regularly held services of worship on Friday evenings, and even though no rabbi was aboard to lead the services, these men have carried on in the tradition of their fathers, using the prayers and hymns of Israel.

A custom which has been popular with the men of the *Missouri* has been that of Evening Prayer, held each evening at sundown when the ship is at sea. Alternating daily between the chaplains on board, the word is passed over the loud speakers: "Stand by for evening prayer!" Throughout the ship men of all faiths bow reverently for a moment while the chaplains leads a brief prayer. Through all the days of fighting and cruising this moment of prayer at the close of the day has brought encouragement and hope to the men of the *Missouri*. The words of the Psalmist have a deeper meaning now than ever before: "They that go down to the sea in ships, that do business in great waters; these see the works of the Lord and his wonders in the deep!"

Prayer At The Surrender Ceremonies

The Missourian Navy Day

October 27, 1945

Eternal God, Father of all living, we offer our sincere prayer of Thanksgiving to Thee on this day which we now dedicate to peace among the nations, remembering another Sabbath Day that was desecrated by the beginning of this brutal war. We are thankful that those who have loved peace have been rewarded with victory over those who have loved war. May it ever be so.

On this day of deliverance we pray for those who through long years have been imprisoned, destitute, sick and forsaken. Heal their bodies and their spirits, O God, for their wounds are grievous and deep. May the scars which they bear remind us that victory is not without cost and peace is not without price. May we never forget those who have paid the cost of our victory and peace.

On this day of surrender we turn hopefully from war to peace, from destroying to building, from killing to saving. But peace without justice we know is hopeless, and justice without mercy Thou wilt surely despise. Help us therefore, O God, to do justice and to love mercy and to walk humbly before Thee.

We pray for Thy servant, the President of the United States, and for the leaders of all lands that they may be endowed with wisdom sufficient for their great tasks. Grant unto all the peoples of the earth knowledge of Thee, with courage and faith to abide within the shelter of Thy sovereign law. Amen.

Sidelight On Surrender

The Missourian Navy Day October 27,

1945

Perhaps the most dramatic aspect of the entire surrender ceremonies lay in the fact that men who for years had been driven and beaten by Japanese guards were permitted to witness the abject surrender of the government and people that had imprisoned them. Heading the list of those present was General Jonathan M. Wainwright, the great leader of Bataan and Corregidor, whose brave leadership in the tragic early days of the war gave the United Stated time to prepare for fighting.

Missouri Gets A "Well Done"

The Missourian Navy Day October 27,

1945

Although the *Missouri* played host to many world-famous and distinguished men, the ceremonies could not have been carried off with precision had it not been for the contribution made by *Missouri* personnel, officers and men. The ship itself had to be prepared for the big event, and as is customary, field day was held for days in advance. Paint was scraped off and fresh paint added where needed, platforms built and a host of other things done to make ready. The Band and Marine Guard of Honor were rehearsed so that their evolution would proceed with clocklike precision. Officer Escorts for all visiting dignitaries were instructed in their duties, under the direction of Commander H.V. Bird, USN. A complete schedule had to be worked out with the Third Fleet Staff so that every phase of the surrender was perfectly timed. The operation of small boats was no small part of the day's activities.

The success of the efforts of the personnel of the *Missouri* was revealed in the congratulatory messages from General MacArthur and Admiral Nimitz who praised the efficiency with which the entire day's proceedings were carried off. To all hands went "A Well Done."

As I Recall

by Henry C. Rivers

My assignment as a platoon sergeant in the Third Fleet Landing Force occupying Yokosuka Naval Base is unquestionably the most unique chapter of the five-plus years I served in the USN. Volunteers with small-arms experience had been requested for this duty. The 15 days beginning August 20, 1945 when *Missouri* Company transferred via cargo nets to USS *Iowa* until September 4 when we returned to a rousing "welcome home" from our shipmates and the band playing, "Hail, Hail, The Gang's All Here" is an experience few can claim.

I recall vividly the sensation of moving between the battlewagons with helmet, Thompson machine gun, 45 caliber pistol, ammunition and full backpack with straps open in front so that it could be freed in case of a fall from the net. I'm quite certain the gear weighed as much as I did in those youthful days.

On August 21 we transferred from the Iowa to USS *Monitor* for nine days of preparation for the ensuing occupation task. Endless drills and instruction in proper use of small arms filled the days. Men grew impatient with the slow progress in chow lines. My free evenings were spent in a space in one of the engine rooms playing deadhand pinochle with a gunner's mate from the 10th Division on *Missouri*, games which often lasted all night. I recognize this man in a picture, but his name escapes my memory.

We moved ashore to Green Beach on August 30 expecting to find a Marine Regiment already there. Only one battalion had arrived. Recollections on arrival include apprehension on landing, a sandy beach, the overwhelming stench of sewage and rotting garbage, bombed-out buildings, and large, appetizing red tomatoes on healthy appearing vines growing on the site of a former building. Consumption of these vegetables was strictly forbidden by the medical officer.

Our quarters were in the Gunnery School Barracks which appeared to be clean, but met our specs only after a regulation Navy field day. This building showed signs of strafing, bullet holes in the end wall and roof from planes which had flown passes over a mountain to the rear. We slept in Japanese hammocks which were stored in the building. One hammock had been left, fully rigged, as if it were a model to follow. Personal belongings were wrapped in our ponchos on the hammocks during working hours.

Toilets were placed at floor level in the barracks. We found it extremely awkward to utilize these facilities but managed until the carpenters built one-holers which were placed over them.

At the outset, we relied on rations and water we had brought with us until we discovered and raided the Marine supplies on the beach. At night, we took our pilfered rations further down the beach, built fires and cooked them. This sustenance satisfied our appetites until "Odom's Stew Palace" was open for business. One night, we walked through a long tunnel in the mountain to the pier on the opposite side of the base to board USS *Dixie* and enjoy a hot meal. We ate in shifts, relieving the watch to allow all men the same luxury.

Nighttime on our side of the base was dark. I set the watch, escorted the men to their locations, and visited them during their watch. Much of our assigned duty was at entrances to caves with fully equipped machine shops and living quarters. The caves were reportedly deserted. Scary duty, the men were constantly apprehensive during the nighttime hours. In the black of night, on approach and identification, the sound of a rifle bolt released was a chilling experience.

There was a warehouse for the personal belongings of the kamikaze pilots. All personal effects had been stashed in wooden boxes, identified with the name of next-of-kin to whom they would eventually be returned. Unfortunately, contents became "souvenirs" for many of the occupation forces. I regret my participation in this activity. As I look at the few pictures of a suicide pilot with his family, his small Japanese flag and decorative fans that I brought home, I have often wished I could find a way to return them to their rightful owners. All of us, winners and losers alike, had been willing to do whatever was necessary for our country.

I found a pair of binoculars in a Japanese cruiser 8-inch gun turret which had been set in concrete on the base. If I still had them, there would be no regret in ownership. Someone else "found" the binoculars in my poncho. I wonder what story has been told time and again about their acquisition.

A small number of unarmed, uniformed Japanese men were on the base. My wife has asked if they appeared frightened or just resigned during the time I was there. I cannot recall any specific emotion or reaction. We did not venture beyond the perimeter of Yokosuka. Just outside the base, I can visualize an arched bridge over a small stream and Japanese women walking with parasols. They quickly lowered them to cover their faces as they passed by.

The only edition of *The Yokosuka Yodeler* was published with this dateline:

Vol. I, No. I, Labor Day, September 3, 1945, Price 1 Yen (Charge it)

Publisher Commander Louis T. Malone, USN, USS *Missouri*, Regimental Commander made this statement: "The circulation of the *Yokosuka Yodeler* is limited to 1,000 and is not entered in the US mails as fourth class mail. The paper is printed in the Yokosuka Navy Yard."

The following announcement appears on Page 1 of 3:

"This is the first edition of *The Yokosuka Yodeler*, the written word of the Third Fleet Naval Landing Force, published and written by the officers and men and the first American publication printed in Japan. Without the benefit of any of the world-famous news services, we have put forth the first foreign military publication ever printed on the Japanese mainland. Frankly, the future of this paper is not bright. We have accomplished our military mission and the situation is well in hand. This is probably the only American newspaper that has been printed on Labor Day. Unfortunately of late, we have been unable to observe a 40-hour week due to extracurricular activities. However, on this point the future is bright. The mimeograph, paper, ink, etc. used are all Japanese prizes of war. For the latest scuttlebutt, turn the pages slowly."

Inevitably, the rumor mill had it that the fleet was leaving without us. After a period of unnecessary uneasiness, we were told that our duty at Yokosuka was over and we would return to our ships. We lined up for our "spoils of war," the choice of either a helmet or a Japanese rifle. I chose a rifle. As with the binoculars, it also was "found" by someone in the armory aboard *Missouri* where it was stored for me on return from Yokosuka. Somewhere in the USA, a former shipmate displays my trophy and tells a tall tale.

The signing of the instrument of surrender on September 2, 1945 aboard the USS *Missouri* was a proud moment for our country and our ship. Countless times in the years since then, I have been asked if I "saw it." No, I did not. My personal proud moment was qualifying as a volunteer for the Third Fleet Landing Force, and for a short time being a participant in the occupation of a vanquished enemy country.

USS Missouri Strikes Again

Submitted by William F. Mowder

With the USS *Missouri* in Korean Waters, October 25, 1952 (our first day) - (NPR): The battleship USS *Missouri* today opened her second tour of duty in the Korean theater with dual-phase gun attacks on a series of eight targets about 140 miles north of the 38th Parallel.

Striking with 16" guns in the morning and 16" and 5" guns in the afternoon, the *Missouri* scored over 20 hits near Chaho and Tanchon, an area on the main east Korean coast supply and rail route.

This *Missouri* action marked the first time that a United States battleship had been called into the Korean conflict for her second tour of duty. The 58,000 ton battleship relieved the USS *Iowa* as flagship of Commander of the Seventh Fleet earlier this month.

The *Missouri* received no enemy fire. However, the ship's 'copter, engaged in gunnery spotting with three officer on board, was fired upon but not hit by North Korean heavy machine guns.

As a result of today's firing, the *Missouri*, under the command of Captain Warner R. Edsall of 2895 South Abingdon Street, Arlington, Virginia, damaged a tunnel mouth with three direct hits, partially blocked another tunnel, cut railroad tracks and started fires in a power plant transformer and a Tanchon factory.

"BATTLESHIP MISSOURI LEFT PEARL HARBOR WITH A 17 GUN SALUTE
APRIL 7, 1951"

"BATTLESHIP MISSOURI LEFT PEARL HARBOR, HAWAII"

USS Missouri Veterans

STANLEY L. ADAMS, born July 4, 1927 in East Syracuse, NY, lived in Frankfort, NY until enlistment in the USN, July 1944. Attended boot training at Sampson, NY and was assigned to the *Missouri's* "K" Div. October 1944. Served as S1/c radar operator, division yeoman and ammunition handler on Quad 16 (during action).

Most memorable experiences: the relentless bombardment of Okinawa, Hokkaido's Steel Works, entry into Tokyo Harbor under a tremendous air armada and his attendance at the WWII signing of the surrender of Japan to the Allied Powers. Signed over for tour of duty to Turkey. Discharged June 1946. Received three Battle Stars with Asiatic-Pacific Ribbon, Victory Medal and other honors issued.

Adams is married to Alice Young (S1/c WAVES). They have two daughters, one son, eight grandchildren and seven great-grandchildren. They celebrated their 50th anniversary Aug. 31, 1996.

He worked for General Electric, Philco Ford and is currently employed at Resorts International Casino, Atlantic City, NJ and is living in Williamstown, NJ.

MATTHEW L. AKERS, born Oct. 26, 1930 in Boston, MA, enlisted in the USN in 1948 and was transferred to USMC July 5, 1949. Served with Marine Detachment on *Big Mo* 1949-50. He has sons, Christopher G., and Geoffrey L. and granddaughter Lily Anne. Retired in 1995 after 35 years on Wall Street to Water Mill, NY on the east end of Long Island.

Served as Marine Orderly to Captains Smith, Duke and Brown. Memories: The morning of Jan. 17, 1949 when Capt. Duke gave that infamous order to the helmsman and in a matter of seconds they were high and dry on a mud bank. And then there was that trip from Norfolk to the Panama Canal in August 1950 heading for Korea. Out of the harbor off Cape Hatteras into a hurricane coming up the coast at 120 mph. What a mess, they were in the canal that next morning.

He took courses through Armed Forces Institute while aboard *Big Mo* and received four year college equivalency degree, left BB-63 in September 1950 with appointment to US Naval Academy, didn't like the prison walls, reassigned to Sixth Fleet with 6th Marines, discharged September 1953 and entered the University of Virginia, Class of 1956, Charlottesville, VA. The rest is history. Semper Fi!

ARTHUR C. ALBERT, born Jan. 29, 1927 in Syracuse, NY, was inducted into the service Aug. 2, 1944 and assigned to the USS *Missouri* September 1944. Plankowner 44-47. At 17 he thought that firemen on ships fought fires. Machinist mate in #3 fire room along with "Hook" Bob Schwenk, finder of straying cases of ham and any unsecured food.

In 1945 a kamikaze hit the *Missouri* slamming Art's knees into a ladder. He kept his injury to himself so he wouldn't distract from the tragic loss of the lives of his shipmates. Fifteen surgeries later he and the *Missouri* still bear the scars.

Albert is such a piece of history in himself that his stepson took him to school for show and tell. When on a vacation to see the huge redwoods, his wife touched the tree and said, "Can you imagine this tree was here at time of Christ!" Then without thinking she reached over and touched Art!

He was awarded the Good Conduct (5), European Ribbon, Asiatic-Pacific, Philippine Ribbon and the American Defense. He was discharged Nov. 20, 1965.

His memorable experience includes taking President Truman to South America with wife and daughter in 1946.

He was married to Martha McIntosh for 34 years. She is now deceased. He is the father of two sons, Al and Larry. He married Sherry Johnson in 1995 and is stepfather to Rhett and Shane. Albert was 1995 Hattiesburg Veteran of the Year. He lives in Hattiesburg, MS, having volunteered for 30 years with the Dixie baseball program for school age boys.

ANTHONY E. (TONY) ALESSANDRO, served on the USS *Missouri* (BB-63) as a seaman first class from 1944 to 1946. His general quarters duties were in the third turret as a powder hoist operator. They fired the 16" projectiles.

Earned four medals: American Area Campaign, Asiatic-Pacific Campaign w/3 Battle Stars, American Victory Medal and the Vaprox Occupation Service Medal.

He had many memorable experiences: on working party with two other shipmates from the 3rd Div., Donofree and Davoli; they brought the Japanese Emissaries aboard from another ship while under way before going into Sagami Wan Bay, they told them about the mine fields and harbor defenses. Entered Sagami Wan Bay, Aug. 27, 1945; passed into Tokyo Bay the morning of Aug. 29, 1945 and dropped anchor in Berth F-71; met his childhood buddy, Frankie Bachanov, from Cincinnati, OH. He was on a destroyer escort which was traveling parallel with the *Mighty Mo* about 200 feet apart going into Sagami Wan Bay when he yelled for Alessandro. He was shocked, what's the chances of meeting your buddy 7,500 miles from home. They met again one year later in Cincinnati and rehashed old times; September 2, 1945, watching from the fifth deck looking down on the surrender ceremony, high ranking military officials of all the allied powers being present. At 0902, General of the Army MacArthur stepped before a battery of microphones and the 23 minutes ceremony was broadcast to the waiting world. Then allied and two Japanese officials placed their signatures on the Surrender Document, MacArthur used five pens signing his name. By 0938, the Japanese Emissaries had departed the ship.

Most recent memorable experience was having the honor to sail the last active battleship in the world, this was December 1991. It was an honor and pleasure to meet the sailors, marines and officer of the *Missouri*. He would like to thank the Skipper Captain Lee Kaiss and the Executive Officer Capt. Ken Jordan. This trip incorporated many memories; the war

was finally over, there would be no more killing and no more dying. They would be going home to their loved ones. He thanks God for having outstanding officers and noncommissioned officers who led their fighting men and women to victory.

Their victory meant freedom from fear and want also freedom of speech and religion. "God Bless America"

ELWOOD E. (ALEX) ALEXANDERSON, born March 28, 1931 in Hammond, IN, joined the USN May 8, 1951. Following recruit training and Class A Electronics Technician School at Great Lakes, IL, he reported aboard *Missouri* on Aug. 6, 1952. On board he served in T-Div. maintaining Mark 25 Fire Control Systems, Aircraft Early Warning System, and various radar repeaters. He attained the rank of electronics technician second class. He served on *Missouri* through Jan. 26, 1955. His most memorable experiences were the 1952-1953 cruise to Korea and readying *Missouri* for decommissioning. Alex was discharged Feb. 8, 1955 in Seattle, WA.

Today, he resides in Albuquerque, NM with his wife, Mary. He is retired after 35 1/2 years as a technical staff member with Sandia National Laboratories, working in advanced system's research and atomic weapon design. Alex and Mary have two children, a daughter and a son, and three grandchildren.

THOMAS I. ALLEN, born Sept. 14, 1928 in Delray Beach, FL, joined the USN Sept. 28, 1950 and assigned to the USS *Missouri* the summer of 1951. He was in the band and achieved the rank of MU2.

While at sea he participated in 26 shore bombardments of Korea and was stationed on the bridge during GQ.

His memorable experiences: At GQ the first time the 16s went off, WOW! Weekend at Ground Zero in Nagasaki and touring the city.

He was discharged July 22, 1954 and received the Good Conduct, China Service, Occupation, Korean Service w/2 Stars, United Nations, National Defense and the Korean Presidential Citation.

Allen retired from Delray Post Office in 1983 and moved to Southport, NC in 1996. He married Althea McHenry in 1955.

FRANCIS J.E. AMPTHOR, born May 28, 1924, was inducted into the USN July 1, 1943 and assigned to USS *Missouri* April 1945. He was discharged Aug. 18, 1946. Educated under V-12 program at Mount St. Mary's College, MD. Graduated and commissioned Duke University on Feb. 23, 1945 with AB in chemistry.

Memorable experiences: member of Gunnery Div. on port side 5" mounts; present at Japanese surrender; brought first signing table from British flagship, found to be too small. Memories include living in "Boy's Town" section of ship; OOD when delivering body of Turkish ambassador home after the war; post war tour of Mediterranean.

Ampthor received the WWII Victory Medal,

American Theatre Ribbon and Asiatic-Pacific The-atre Ribbon w/2 stars.

He married Dorothy Carson and they have sons, Robert and Blane; daughters, Irene Tori and Lois Burns and seven grandchildren.

He retired from Rohm and Hass Company in 1990 after 44 years as research chemist and taught at St. Joseph's University (Phil) night school. He was active in church, community, USS Missouri Association, His sons are now honorary members.

RALPH ARONE, born May 4, 1920 in White Plains, NY, joined the USN in 1948 and assigned to USS *Missouri* in 1949. He served as electrician and achieved the rank of PO3. While at sea he served in the Korean War.

His memorable experiences: while leaving port in Norfolk, VA to go on a cruise with midshipmen the ship ran aground.

He was discharged in 1952 and received the Korean Service Medal and the Navy Occupation Service Medal.

Arone retired from electrical union after 38 years. He is married and has one son, one married daughter and four grandchildren.

TROY A. ARRINGTON, born Aug. 6, 1969, Colville, WA, enlisted in the USN June 15, 1988. Stations include: San Diego, CA; NTC Great Lakes, IL; USS *Missouri*, USS *Tarawa*. He was discharged November 1944.

Served aboard USS *Missouri* from April 1990 to March 1992 GM Div. - technician/operator Phalanx MK-15 MOD 0 20 mm gatling guns.

As a FC-3 he remembers being shot at by two Silkworm missiles during the Persian Gulf War (Desert Storm) and shaking hands with General Colin Powell during the 50th anniversary of the Pearl Harbor attack.

While aboard Arrington received Combat Action Ribbon, Navy Unit Commendation, National Defense Medal, Southwest Asia Service Medal, Kuwait Liberation Medal and Battle "E". Also received Letter of Commendation from the Admiral and was nominated for Sailor of the Year, at which time he was given the honor in meeting and shaking hands with former President Ronald Reagan.

Currently resides in Costa Mesa, CA with his wife, Tammy and four year old daughter, Ashley Marie. Employed as a CNC technician at Dynamotion in Santa Ana, CA.

RICHARD C. ARRUDA, born Sept. 25, 1925 in Fall River, MA, entered the USN Sept. 25, 1943. Stations include: NTS, Newport, RI. Radio training at Boston Radio School. Assigned and plankowner on USS *Missouri* until discharged March 1946 in Brooklyn, NY.

He was a member of the CR Div. (Com. Radio) with rank R3/c. Duty: Radio 1 code, on bridge w/ ship to ship communication. Battle station conning tower, and responsible for corrections and changes in radio call sign book of all ships and stations.

He received the American Theatre and Asiatic-Pacific Medal w/3 stars.

He is married to Vivian Nadow, and they have a son and daughter, 10 grandchildren and five great-grandchildren. Arruda retired to the golf course and resides in Attleboro, MA.

ROBERT L. BALFOUR, born in Wisconsin May 16, graduated from Coloma, Michigan High School in 1935, two years of college at Kalamazoo College and received a BU degree from the University of Missouri in 1940 as class president. He became an Eagle Scout in 1933.

Following graduation Balfour was a newspaperman in Iowa and Michigan, general manager of a new radio station, an officer in an advertising agency, assistant to the president of a publishing corporation and served in the USN 1943-1946 as an officer on Adm. Halseys Commander South Pacific and Commander Third Fleet staffs. He served again in 1950-1951 during the Korean War.

Balfour was campaign manager for his USN shipmate Gov. Harold Stassen in his 1952 bid for the presidency and later travelled 51,000 miles through 44 states as a member of President Dwight M. Eisenhower's pre-election campaign staff. He served as a VP and sales manager for Purex, Calgon and Johns-Manville subsidiaries and VP and general sales manager for the Club Car Golf Car Company in Augusta, GA.

Balfour and Stassen are the only two to have served on the staff of both a five-star General (Eisenhower) and a five-star Adm. (Halsey).

Sam Riley, professor of Communications Studies at Virginia Tech, recently published a book, *Biographical Dictionary of American Newspaper Columnists,* offering his choice for the 600 best newspaper columnists from the time men and women having this job description appeared in US newspapers from the Civil War to the present.

Among the 600 Hall of Famers are such eminent writers as Lewis Grizzard, Edgar Guest, Ernie Pyle, Mike Royko, Irvin Cobb, Carl Sandburg, Walter Winchell, George Will and LCDR Robert Balfour, who as a CWO on Adm. William Halsey's commander 3rd Fleet staff was aboard Halsey's flagship, the USS *Missouri,* in Tokyo Bay Sept. 2, 1945 at the formal surrender of the Japanese forces.

Balfour, the oldest member of the National Society of Newspaper Columnists at the age of 79, has had a book, 38 national magazine articles and 331 newspaper columns published in Georgia and Florida newspapers and on a wire service since retirement.

LANDER F. BARKER JR., born April 11, 1928, Gastonia, NC, enlisted in the USN Feb. 13, 1946. His stations include: NTS Norfolk, VA, USS *Missouri* and NAS Norfolk, VA.

He was assigned to USS *Missouri* in May 1946. In November 1946 they took the *Missouri* on cold weather shake down cruise to Arctic Circle.

His most memorable experience was being hit by a star shell from the USS *Little Rock* while on night time exercise, one man was killed.

Barker was discharged at NAS Norfolk, VA on Dec. 11, 1947 as S1c. He is presently working for Clerk of Circuit Court, Indian River County, FL (Vero Beach). He has two sons, Lanny and Brent and two grandchildren, Janine and Carissa.

JAMES M. BARNWELL, born July 6, 1925 in Hendersonville, NC, was inducted into the USN Sept. 9, 1943 and assigned to the USS *Missouri* 1944-1945. He served in 3rd Div. as assistant gun captain, right gun turret 3, seaman first class.

His memorable experiences include being plankowner on *Missouri*; winning Army-Navy E for 16 inch turret.

While at sea he participated in the invasion of Iwo Jima, Guam and Okinawa. He was awarded the Asiatic-Pacific w/3 Stars, American Theatre, WWII Victory and National Defense Service Medal. Discharged Dec. 27, 1945 as SFC.

He is one of seven brothers who were in service during WWII. A retired SFC US Army, Civil Service and SS.

JOHNNIE M. BARR, born Sept. 6, 1920 at Mt. Solan, VA, enlisted in the USMC Aug. 29, 1942 and assigned guard duty in Washington, DC and later to the USS *Missouri* during its construction and commission.

He served on *Missouri* until discharged Oct. 26, 1945 and had the pleasure of being present at the commissioning the signing of surrender; last decommissioning at Long Beach, CA March 31, 1992 and at the 50th anniversary at Bremerton, WA Sept. 2, 1995.

He married Neva Elizabeth Stroop on Feb. 26, 1944 and they are still together after 53 years, Feb. 26, 1977. He has one son and one daughter. Johnnie Barr Jr. married Teresa Horner and they have two daughters, Lindsey and Amanda. Their daughter, Teresa married Barry Shifflett and they have a daughter, Lisa and a son, Brian.

Barr retired from the traffic department of the trucking industry in 1985 after 32 years. He enjoys fishing.

ORLANDO D. BATISTA, born Aug. 13, 1933 in Ludlow, MA was inducted into the USN on June 26, 1951 and reported for basic training at the Great Lakes NTS. Upon completion of basic he was assigned to the USS *Missouri* in Norfolk, VA. He served as a fireman in the "A" Div. as a refrigeration and aircondition repairman. Completing a shakedown cruise to Europe the *Mo* then sailed for Japan for the Korean conflict, the second tour of duty in those waters. He was assigned to submarine training at New London Submarine Base in December for a short stint and later to the USS *Monssen* at Newport, RI.

His awards include the National Defense Service Medal, United Nations Service Medal, Korean Service Medal, ROC Presidential Unit Citation and the China Service Medal. He was honorably discharged June 25, 1959.

He left the service June of 1954 and married Phillis Guyette in August of 1956, together they had three daughters, Lori, Lisa and Tina. As life went on they also became grandparents of five girls and two boys. Orlando retired in 1992 from the Monsanto Company of Indian Orchard, MA after 36 years. He retired as a maintenance supervisor and the couple now spends the winter months in Largo, FL.

VERNON J. BAYS, born Feb. 21, 1923 in Eccles, WV, joined the USN Dec. 20, 1940 and was assigned to USS *Missouri* Aug. 4, 1950 in damage control, achieving the rank of damage control mate first class.

While at sea the ship's location was between North and South Korea at the 38th Parallel. The ship's mission was to bomb railways, roads, and bridges and numerous ports: Seoul and Kunsan on the east coast of the Yellow Sea and Samchok and Yongdok on the Sea of Japan.

One of his proudest moments was receiving a commendation, one of 223 out of 2,000 men, from Captain Duke for air testing the living compartments and voids of the ship in one year's time when the assignment would normally take three years. This assignment was necessary since damage reports were left undone for the time the ship was aground. In that emergency, items like storage lockers, 5,000 pairs of combat boots and 400 life jackets that were discovered on the bow were taken to the stern to bring the bow out of the mud bank. The reports had to be completed so the ship would be combat ready to continue on to Korea. Water-tight integrity was always of the utmost importance for safety.

Captain Duke hand delivered the commendation to him in sick bay as he was healing from an accident where a 2,000 pound hatch cover dropped on his foot as he was freeing six gunners mates who were trapped under #2 turret three decks below in an ammunition storage area.

One of the scariest times was going through a typhoon off the Carolina coast on the way to Korea. Their ship, nor any other, should go through this type of dangerous weather and storm conditions. During this stormy trip his mind flashed back on an incident that made naval history - a flashback of a rather sad moment of making a packing box for the personal effects of Admiral Wilcox, the only Admiral to have plunged to his death in an accident at sea while aboard the USS *Washington* in March of 1942. Like the waves in that storm, as a young man of 23, he had already experienced the turbulent ups and downs of life in the USN aboard a battleship.

He thinks the most humorous time was receiving a package from his wife - it was pint bottle of "Four Roses" delivered by destroyer mail on Dec. 24, 1950. He'll always wonder how it made it half way around the world in one mail sack after another and reached him, intact, on Christmas eve.

QUENTIN F. BEATY, born Sept. 14, 1922 in Jamestown, TN was inducted in the USN Aug. 11, 1943. He was stationed at Bainbridge, MD, Norfolk, VA and Newport, RI where the original crew was formed.

A plankowner, he stayed aboard and was a part of the formal surrender on Sept. 2, 1945 and was discharged at Memphis, TN on March 13, 1946 as S1c with the American Campaign Medal, Asiatic-Pacific Campaign Medal, WWII Victory Medal and the Navy Occupation Service Medal w/Asia Clasp.

He will never forget seeing the aircraft carrier, *Franklin,* get hit by a bomb one morning about sunup and the time a Japanese suicide plane come in on the starboard side and crashed on the ship and left a machine gun barrel from the wing of his plane stuck in the muzzle of their 40mm quad. But being present at the surrender ceremonies is his most valued memory.

The little book of history and pictures was given to all of the men and he is most proud of pictures of he and a little string band playing on deck entertaining their buddies. He was the guy with the (fiddle) violin.

Beaty is married to June Thomas and they have seven children. He retired after 35 years as an electrician living in Muncie, IN.

LLOYD WILBUR BEEBY SR., born Oct. 30, 1925 in Philadelphia, PA, was inducted into the USN Oct. 13, 1943 and assigned to the USS *Missouri* Dec. 23, 1949. He was stationed at 40mm Btry., Std. Side, 7th Div.

His memorable experiences include being in gunners mate school when the ship ran aground.

He served in action in Korea and WWII. His awards include the American Area, Good Conduct Medal, Asiatic-Pacific w/4 stars, Philippine Liberation w/2 stars and WWII Victory Medal.

Beeby was discharged April 15, 1963. He is married with two sons and one daughter. He retired after 28 years working in Kennedy Space Center, FL. He has been owner of a golf shop for 23 years, making clubs and selling all golf equipment.

D.W. BISHOP JR., born May 21, 1919 in Little Rock, AR, and raised in Kansas City, MO, attended three years of college. He received his commission into the USN December 1940.

He served on the USS *Texas* from Spring 1941 until February 1944. While on the *Texas* was 2nd Div., Main Btry. officer and officer of deck. While on *Texas* participated in North Atlantic patrol and North African invasion. February 1944 transferred to USS *Missouri* for pre-commissioning and commissioning. He was 1st Div., main battery officer and officer of deck underway.

Bishop served on USS *Missouri* from her commissioning until October 1945, then returned to inactive duty with the rank of lieutenant commander. Participated in invasion of Iwo Jima, Okinawa, mass air strikes against Japan, bombardment of Japan and the signing of Japanese surrender on the USS *Missouri*.

After the war he entered the furniture manufacturing business with his father. Retired from furniture business in Lebanon, MO 1984. Served on the Civilian Liaison Committee for the recommissioning of the USS *Missouri* in 1986.

He married Gwendolyn B. Fleming August 1942 and has two sons, D.W. (Bill) Bishop III and Richard D. Bishop, three grandchildren: Jarrod, Drew and Hannah.

DAVID M. BLACK, born Oct. 23, 1924 in Winburne, PA, enlisted in the USNRS-SV on June 8, 1943, immediately after high school graduation. His stations include: basic training - Sampson Naval Station; NTS NOB Norfolk, VA; NTS Newport, RI; Brooklyn, NY where the USS *Missouri* was commissioned. Served on the USS *Missouri* until discharged March 12, 1946 from Sampson, NY.

Seaman First Class Black received the Victory Medal, American Campaign Medal and Asiatic-Pacific Medal w/3 stars. He is a member of the 11th Div.

His most memorable event was assignment to the Blue Jacket Landing Party, Task Force 38, which was made up entirely of naval personnel and was the first landing party at Yokosuka Naval Yard, Japan. They secured the base and remained there during the signing of the peace treaty on the *Missouri*.

Black married Beverly Bowdler (deceased). He is the father of five children. He later married Harriet Rollins-Robertson and is the stepfather of two children. He retired from the Niagara Falls New York Fire Department after 33 years. He and his wife reside in Rushford, NY.

DONALD H. BLOCH, born July 27, 1926 in Reading, PA, enlisted in USNR Jan. 19, 1944. He received a high school education. Served on board USS *Mercury* (AKA-42) during WWII, participating in the invasion of Saipan Island, Mariannas, Leyte Island and Luzon Island, Philippines and in logistic support of Task Force 58 during Okinawa campaign.

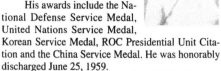

His most significant memory of WWII was being torpedoed during Marianas campaign. He was honorably discharged April 6, 1946 as QM3.

Enlisted in USN as SN, Feb. 17, 1948. Reported on board USS *Missouri* for duty March 5, 1948. While on board *Missouri,* advanced in rate from QMSN to QM1; received CO's commendation while in Korea. His tour on board *Missouri* was unique in that he experienced her grounding and shame and to her providing ground-fire support during the evacuation at Hungnam, Korea, putting her back in her majestic glory again.

He was transferred to USS *Columbus* (CA-74) on Feb. 18, 1952 and had a tour of duty as company commander, USNTC Bainbridge, MD; tours of duty aboard USS *Murray* (DD-576); USS *Galveston* (CLG-3); and the Pentagon, Washington, DC, where he ended his USN career by retiring as QMC, with 21 years service.

He then went to work on the Reading Railroad/Conrail, finishing his second career as locomotive inspector, on Aug. 1, 1988.

Next to marrying Betty Kelchner and having a son, Donald Jr., and acquiring three great-stepchildren: Larry, Barry and Linda, his tour aboard *Missouri* was the high point of his life.

AMBROSE L. BOHAGER, born Aug. 22, 1922 in Baltimore, MD, entered the USN Oct. 26, 1942. His stations include: NRS Baltimore, MD; NTS Norfolk, VA; USNTS Bainbridge, MD; USNTS Newport, RI; served on USS *Missouri* from June 1944 until discharge Nov. 13, 1945.

His memorable experiences include the crashing of a kamikaze plane on April 11, 1945 at Okinawa into the ship's starboard quarter, just below the 40mm anti-aircraft gun he was stationed at. His rank was musician third class.

His awards include the American Theatre, Asiatic-Pacific Area w/3 stars, Victory Medal, American Campaign Medal and Asiatic-Pacific Campaign Medal.

As a USN band trumpeter, he played "God Save The King" and "The Star-Spangled Banner" at the surrender ceremonies aboard the ship Sept. 2, 1945.

After the war, he became Clerk of the Superior Court in Baltimore City, retiring at age 63. He also continued playing trumpet and remained a well-known local musician until his death in 1987.

Surviving are his wife, Wilhelmina (Hiebler) Bohager, and three sons: Leonard, David and Robert. *Respectfully submitted by Leonard Thomas Bohager, honorary member USS Missouri Association, Inc.*

NELSON D. BOOKSTAVER, born in NYC, April 12, 1924. Entered USN service: V-12 Unit 1943-1945, reserve status until 1948. Active duty 1948-1949, USS *Missouri*. Recalled to active duty 1952-1954, Brooklyn Navy Yard, then Boston, Receiving Station (Fargo Building).

On *Missouri* completed two midshipman cruises (England and France). Panama cruise, New York, Guantanamo Bay.

He married in 1954 to Miriam Synes and they have two daughters. Completed post graduate training in orthodontia, practiced in Bergenfield, NJ and is now semi-retired.

DOYLE E. BOOTHROY, born March 28, 1926 in Mitchellville, IA, enlisted in the USN on Sept. 16, 1943. He attended boot camp at Farragut, ID; Aerology School at Lakehurst, NJ and had numerous other duty stations in the US.

May 8, 1945 he was assigned to Adm. Halsey's Third Fleet Flag Allowance as a weather observer AERM3C. May 18, 1945 Halsey and his flag boarded his flag ship, the USS *Missouri* (BB-63) in Guam.

His most memorable experience was witnessing the signing of the formal surrender of the Japanese which ended WWII Sept. 2, 1945.

Doyle was discharged March 18, 1947. Today he is retired and lives in Ocala, FL with his wife, Margie.

They are celebrating their 50th wedding anniversary Nov. 8, 1997. They have a son Cranson, one daughter, Candance and two granddaughters, Rachel and Megan Bower.

VICTOR J. BOSCHINI SR., born July 5, 1922, joined the USN Aug. 4, 1941 and assigned to the USS *Missouri* Jan. 19, 1944 in hyd. and machine shop as MM1c.

His awards include the Navy Unit Commendation, Good Conduct, American Defense, American Campaign, Asiatic-Pacific Campaign, European Campaign, WWII Victory, Navy Occupation, Asia Clasp and Philippine Liberation.

His memorable experiences include: living on the top deck of a barrack across from the navy yard. One night Ralph Carlton MM2c came aboard wearing a mattress cover and a flat hat. The O.D. was fit to be tied. They had to recuse him. The OD was a retired WO. He said he had never seen this before; the trip to Rome and Turkey; audience with Pope Pius XII.

Boschini's license plate has been BB-63 since 1950. He is now retired from the Ford Motor Company and is married to Betty. They have children Gail Ann, Vic and Beth (twins) and seven grandchildren, which includes twin boys.

MICHAEL F.X. BOYLAN, enlisted in the USN August 1944 at 17 years old after graduation from high school. After boot camp at Sampson, NY he was assigned to USS *Missouri*. Served in KE Div. (Radar Repair Div.) as electronic technician's mate third class (ETM3c), 1944-46.

Soon after discharge appointed to Jersey City, New Jersey Police Department. Served for 30 years, attaining rank of captain (communications officer). While in police department attended Cooper Union School of Engineering (New York City) for eight years at night and in 1966 received bachelor of science degree in Electrical Engineering (BSEE). Presently employed by US Army at Fort Monmouth, NY as electronics engineer working primarily in field of helicopter avionics (aviation electronics).

He is married to Edna (Sheridan) and they have four children: Michael, Margaret, James and Thomas and seven grandchildren.

ROBERT C. BRAY, born Dec. 31, 1924 in Kansas City, KS, joined the USN July 6, 1943 at the federal building in Kansas City, MO. His first promotion was right after their group was sworn in. They were all made captain of the head and cleaned the rest room of the building.

In Newport, RI in April of 1944 he was assigned to USS *Missouri* pre-commissioning detail. He boarded *Missouri* in the Brooklyn Navy Yards in June 1944 and was aboard for commissioning. His duties were maintenance on Quad 40mm Mount #2, later on 40mm Mount #6, and gun captain at General Quarters. He recalls going through a typhoon off the coast of Japan. The intensity of the typhoon ripped a bolted down,

20mm ready ammunition box from the main deck, and set it on top of 16" turret #1. Later they bombarded steel mills on the Japanese mainland with their 16" guns, all while Adm. Halsey had his flag aboard the *Missouri*. Bray was discharged March 12, 1946.

FRED G. BRETL, born April 26, 1926, in Schwandorf, Bavaria, came to the US in 1929. Grew up in Buffalo, NY and graduated from St. Ann's Business School in 1942. In October of 1942 he enlisted in the USN. He had to use his friend's birth certificate and enlisted as John Albert Kraus. After serving one year in the USN, this was discovered, but luckily it turned out fine, after proving he only had good intentions and wanted to serve his country.

He served on the *McKenzie*, the *Monitor* and on the *Missouri* from May 1944 to September 1946. He was present for the signing of the surrender of the Japanese. He became a citizen during his assignment on the *Missouri*.

As yeoman second class he received five ribbons, EAME, Asiatic-Pacific, American Theatre, Philippine Liberation and Good Conduct.

Bretl married Maryann Grandits in 1950 and they had three daughters and two sons. He owned a flower shop for 35 years and retired in 1990.

RICHARD BROCCO, born March 18, 1926, Bird-in-Hand, PA, enlisted in USN March 16, 1944 and was assigned to the *Missouri* when commissioned, June 1944. He served on the *Missouri* until discharged May 16, 1946. CR Div.-Communication, seaman first class.

He will never forget seeing the Japanese suicide plane coming in for the hit; witnessing the signing of the Peace Treaty on Sept. 2, 1945.

Brocco received the Pacific Theatre Ribbon (three stars), American Theatre Ribbon and Victory Medals.

He is contractor-builder and very active in the Masonic Fraternity, 33 Degree Mason in the Valley of Lancaster, PA.

He married Mary Frances Ault Brocco Sept. 11, 1948, and they have three daughters: Deborah Ann Tracy, Cynthia Marie Quinn and Kathy Frances Hess and five grandchildren.

MAYNARD G. BULAND, born Feb. 11, 1931 in Albert Lea, MN, was inducted into the USN February 1951 and assigned to the USS *Missouri* in the metal shop and maintained and repaired ventilation ductwork.

His memorable experience includes seeing Korea, Japan, Norway, England, France, Rio de Janerio and Panama.

While at sea he participated in Korea and midshipmen cruises and Buland received the UN Medal, Korean Medal, Good Conduct Medal, National Defense Medal and China Service. He was discharged February 1955 as ME2

He married Norma and has sons, Glen and Eric and a daughter Susan. He is now retired.

GEORGE W. (BILL) BURCH, born Jan. 28, 1928 in Beech Bottom, WV, enlisted when he was 17 years old and after boots at Great Lakes was assigned to the *Missouri*. A member of the Engineering Department, he served in the #2 engine room, generator and shaft alley watch. His GQ station was Repair 5.

His memorable experiences include bringing President Harry Truman back from Rio and having shellback ceremonies with the Trumans; watching President Truman climb the ladder up the side of the #3 turret and greet the surprised fellows that were up there; crossing the Arctic Circle around Thanksgiving 1947, a test of the men and equipment.

He never thought he would ever see the *Missouri* activated again once it was mothballed but when it was recommissioned he and his wife Jean flew out to Long Beach and attended "Old Timers Day." They climbed everywhere, even up to the O level. "That's the tiredest I ever was," he exclaimed.

Burch is retired and his hobbies are woodworking, photography and helping his daughter, Linda with her arts and crafts in Charleston, SC.

DONALD STEWART BURR, born and lived in northern California, mostly in Fresno. Following two years of college, he joined the active Naval Reserve. After the Pearl Harbor attack, the USN put him to work teaching radio technicians for two years. He rose from seaman to CPO during this time. Burr joined the *Missouri* pre-commissioning detail at Brooklyn in April 1944, and stayed as CPO of the Radar maintenance (K) Div. (also known as KE Div.) until April 1945.

After the action near Okinawa described in his story, he was sent stateside for new construction. He married WAVE SK(D)2c Frances Leone Holt, and was sent briefly to Philadelphia for construction of a new carrier, USS *Franklin,* and discharged there.

Burr was called up to the Naval Radio Station, Pearl Harbor (NPM) in 1950, and was assigned to the 500,000 watt transmitter.

He worked about 40 years as an operating engineer at radio and TV broadcast stations in Fresno, CA.

GEORGE (SEAN) BUTLER, born Feb. 17, 1965, enlisted in the USN in June of 1983. After completing MMA School and Nuclear Power School he reported to the *Missouri* on March 18, 1985 as a MMFN. He left the *Missouri* on Jan. 29, 1991 as a MM1 to attend Cryogenic School. From there he spent three years at NAS Agana, Guam and two years aboard the USS *Nimitz*. He is currently stationed at NavSta Everett.

While stationed on the *Missouri* he was assigned

to engine rooms 1 and 2. During his time on the *Mo* he went on a World Cruise, the Persian Gulf Crisis on 86 and Desert Shield/Desert Storm.

Chief Butler has earned three Navy Achievement Medals, National Defense, Naval Expeditionary, Southwest Asian and both Kuwait Liberation Medals.

He is married to Lisa Schopp and has two sons, Brendon and Mackenzie. Chief Butler currently resides in Bremerton, WA.

FREDERICK CANBY, born Jan. 14, 1961, Ft. Benning, GA. Commissioned ensign, Dec. 31, 1985. Assignments: Naval Dental Center, Long Beach, CA two tours and USS *Missouri*, July 1990-March 1992. Six years active duty. Currently, lieutenant commander, Dental Reserve Unit, National Naval Dental Center, Bethesda, MD.

Memories: camaraderie during Operations Desert Shield/Storm; sunrise and sunset in the tropics; meeting the surviving dental officer from the damaged USS *West Virginia* during the 50 year anniversary of Pearl Harbor.

He was awarded the Navy Achievement Medal w/star, Combat Action Ribbon, Navy Unit Commendation, Navy "E" Ribbon, National Defense Medal, SWA w/2 stars, Sea Service Deployment, Kuwait Liberation (Saudi Arabia and Kuwait), Navy Expert Pistol. Surface Warfare Medical Dept. officer.

He received a BA in Colby College; DDS, Georgetown University; MS/Certificate in Endodontics, Northwestern University.

Canby is employed in private practice in Northern Virginia and is married to Denise Canby. They have one daughter, Adrianna and a second child due in November 1997.

CHARLES EMIL CARLSON, born Aug. 21, 1925 and inducted into the USN Aug. 29, 1942 and assigned to the *Missouri* March 11, 1944 as plankowner.

His stations 11th Div. - gun site operator (40mm and 5" guns), changed to SS Div. after Peace Treaty Signing.

His memorable experiences: Peace Treaty Signing and attending re-commissioning of *Missouri* in San Francisco.

Prior to *Missouri*, served on USS *Marblehead* October 1942-September 1943; patrolling for enemy ships and submarines in South Atlantic; served on USS *Quincy* December 1943-March 1944; attended gunnery training in Rhode Island after being assigned to *Missouri*.

He was awarded the Pacific Theatre (three stars), American Theatre, European Theatre and Victory Medal. Carlson was discharged Aug. 19, 1946.

He has been married to Mary Kurz for 52 years and they have three children, six grandchildren, and four great-grandchildren.

Carlson retired after 27 years with city of San Francisco and enjoys visiting his children (two live in Hawaii and one in San Diego). He also enjoys sailing, snowskiing, and kayaking.

ROBERT W. CARMINT JR., born Dec. 10, 1928 in Philadelphia, PA, enlisted in active Naval Air Reserve at NAS Willow Grove, PA, June 1949. Reported to active duty, November 1951 at NAS Pensacola Mainside, FL as "weather guesser" (aerographers mate).

Served aboard the USS *Cabot* (CVL-28), Naval Training Command carrier, conducting carrier qualification landings in Gulf of Mexico. The *Cabot* transferred to Atlantic Fleet, NOB Norfolk, VA. Sailed to North Atlantic (Newfoundland) on cold weather exercise with Marine Helo squadron aboard USS *Albany* (CA-123). Sailed to Caribbean with Marine regiment for maneuvers on Vieques Island and gunnery training.

Flag transferred to USS *Missouri* (BB-63). Picked up midshipmen at Annapolis for training cruise to Rio de Janeiro, South America.

Carmint was honorably discharged, June 22, 1953 as aerographer third class. He is married to Kathleen S. Neely and has one daughter Reneé and four grandchildren: Alexa, Jeremy, Brianna and Jason.

Retired from Prudential Insurance July 1996 and is now enjoying the best years of his life.

DONALD P. CARON, born Sept. 29, 1928 in Westbrook, ME, left high school his freshman year. Enlisted in the USN July 15, 1948, trained in Great Lakes, IL and sent for three weeks training in CIC at Norfolk, VA. His first assignment was on the USS *Missouri* in 1948. He was in the K Div. or radar. On Aug. 25, 1949 he was assigned temp. duty as a French interpreter. He was in Cherbourg, France, doing interpreting for the OOD and other official and social functions. Assigned a proficiency mark of 3.0 in ability to understand and interpret French. His rank was SA.

Memorable experiences include seeing places he had read about in school; aboard when the ship went aground. Caron was discharged Feb. 15, 1950

Married Harriet Le Clair Aug. 12, 1950 and they have been married for 42 years. They had four children, nine grandchildren and one great-grandchild.

He belonged to the Honor Guard of the Missouri Association. He attended the recommissioning.

Retired from Scott Paper Co. after 40 years and passed away March 7, 1993 with the Honor Guard in attendance. He has the Mighty Mo engraved on his monument.

MATTHEW J. CEGELIS JR., born June 3, 1932 in Lawrence, MA, graduated Searles High School, Methuen, MA in 1950 and enlisted in USN August 1950. Served aboard USS *Macon* (CA-132) from 1950

to 1952 and transferred to the USS *Missouri* (BB-63) Aug. 29, 1952. Served aboard the *Missouri* until April 1954.

A lot of things happened and there is a lot to remember. Going through the Panama Canal with inches to spare; gun strikes in Wonson Harbor, Korea; Pearl Harbor, HI; crossing the equator; going to Rio de Janeiro, Brazil and much more.

Cegelis was discharged as IC E3c and received the Good Conduct Medal, National Defense, Occupation Service, China Service, Korea, United Nations and Korean Presidential Award.

He married Christina Palese, has a son, Matthew J. Cegelis III, a daughter Sylvia M. Wallace, a grandson and a granddaughter. Retired from American Trucking Association as regional vice president after 35 years.

RICHARD CHABOT, born May 4, 1928, Wharton, NJ. He was sworn in the USN July 24, 1945 and assigned to the USS *Missouri* May 9, 1946. Duties included main ice machines, air conditioners, water coolers and small refrigerators. Served during peacetime and only action was friendly fire.

Memorable experiences include December 1946 when hit with 5" star shell on leave from USS *Little Rock* and killing our coxswain; loss of reserve pilot off Norfolk when his plane was catapulted off ship and flipped over (Summer 1948); trip to Rio in August 1947; returning with President Truman, wife Bess and daughter Margaret; Brooklyn Navy Yard, 1947-48.

Richard was discharged as MM2/c April 26, 1949. He received all the usual peacetime medals.

Retired, his last full time job was as traffic manager for Conder Corp. Richard has been married since Aug. 9, 1952 and has six sons and one daughter.

EDWARD GENE CHAPMAN, born Aug. 15, 1934 in Auburn, NY, joined the USN Aug. 31, 1951 and was assigned to the USS *Missouri* November 1951, MM B Div.

While at sea he participated in Korea. His memorable experiences include being on smoke watch #1 stack and hearing shells whistling over head.

Chapman was discharged July 31, 1955 as MM3 and received the National Defense Medal, Korean Medal, UN Medal, China Service and two Battle Stars.

He has four sons, one daughter, 11 grandchildren and one great grandchild. He retired from Johnson Controls after 31 years in Owosso, MI. Presently manager of a condo complex on the beach at San Clemente, CA.

ALFRED J. CHARPENTIER, born July 30, 1932 in Fitchburg, MA, joined the USN May 9, 1951 and was assigned to USS *Missouri* September 1951. He served as gunners mate 5th Div. Mount 1.

While at sea he participated in the Korean Conflict. His memorable experiences include Wonson Harbor; Panama Canal; being in Portsmouth England for Coronation of Queen Elizabeth.

His last five months in the USN he served on the *Newport News* and was discharged off the USS *New Jersey*. (Only because he had little time to serve and did not want to be released on the west coast). Had he stayed, the *Big Mo* would have been his only ship.

Charpentier was discharged May 1955 as GMSN. He received all awards the *Missouri* received from 1951-1955, including the Good Conduct Medal.

He married Madeline Canistro in 1953. They have one daughter, Donna and twin sons, Fred and Bill and grandchildren: Gregory, Christopher, Stacie and Andrea Charpentier. Retired in 1994 after 40 years as a tool & die maker and is now a semi-retired tool maker. They are traveling when they can and enjoying life.

RALPH L. CHELF, born Jan. 17, 1926, Knifley, KY, entered USN May 4, 1944. His stations include: boot camp, Great Lakes, IL; Gunnery School, Gulfport, MS; SS *Eloy Alfaro,* a liberty ship delivering war materials in Murmansk, Russia.

He went on board the *Missouri* Jan. 4, 1946, serving through Oct. 5, 1947. He was a coxswain in the 3rd Div., gun captain of a 16" gun in turret 3.

His favorite memories include the 1946 visit to Istanbul, Turkey and other ports of call throughout the Mediterranean; the 1947 visit to Rio de Janeiro, Brazil when the Organization of American States (OAS) came into being. President Truman returned to the US on board the *Missouri* and walked through the ship daily. He would stop and pose for pictures except when wearing swimming trunks.

Chelf was one of several sailors selected to have lunch with the President and his party. He was further honored on Saturday morning at captain's inspection, the President shook hands with him and asked his name, where he was from and what his campaign ribbons signified.

While on board the *Missouri* they crossed the equator enroute to Brazil and the Arctic Circle while operating in Baffin Bay.

He retired from the USN Dec. 23, 1963 as senior chief electronics technician and from Kentucky State Government Jan. 1, 1991.

THOMAS S. CHRISAFIS, born Sept. 26, 1926, was inducted into the USN, March 11, 1944 and assigned to the USS *Missouri* April 30, 1944. His stations included seaman and gunnery duty.

Memorable experiences: Sept. 2, 1945, witnessed the unconditional surrender of Japan. While at sea he participated at Iwo Jima, Okinawa, battle and occupation of Japan.

Chrisafis received the Occupation Service Medal, Philippine Liberation, Asiatic-Pacific Campaign w/3 stars, and WWII Victory. He was discharged May 13, 1946.

He served with the WWII Pacific 3rd, 5th and 7th Fleets.

ALBERT A. CIRIELLO, born Oct. 30, 1926, joined the USN March 28, 1944 when he had just turned 17.

His duties were in the IC Room Forward. The *Mighty Mo* departed from Pearl Harbor on Jan. 1, 1945 bound for the Ulithi Islands in the Carolines. On the way to Ulithi, there was a tapping sound that was not normal to the ship. His station was below the heavy armament and this sound would come into play every evening at about 1600 hours. He and his fellow mates

took turns looking for the tapping sound. After several weeks they found the source. A crew member two decks above was making silver rings out of silver dollars by tapping the edge of the coin with much patience and using the deck for an anvil. The rings were quite good but once the officers got wind of this practice the tapping stopped.

Ciriello was on board the *Missouri* from the day it was commissioned in Brooklyn until he was discharged on May 13, 1946 as an EM2c with three Pacific Stars. He is a plankowner and a plaque owner.

He is now retired from his own company, Marine Propulsion Engineering Company. He and his wife Dorothy have three children: Helene, Anthony and Albert, all married, and one grandson, Shawn.

LESLIE BUFORD (BUDDY) CLANTON, born Nov. 17, 1926 in Quitman, GA, educated and resides in Albany, GA. Enlisted in USN Nov. 15, 1944. Stations include: Great Lakes, IL, Co. 2172; Ft. Pierce and NAS Jacksonville, FL; USS *Vanakis* (AKA-49); Armed Guard Brooklyn, NY; USS *Missouri* May 9 1946 to Sept. 19, 1947. Discharged at Norfolk, VA as GM3c on Oct. 13, 1947. He (WWII) trained at Newport, RI and his father (WWI) also USN.

He married Betty Harris and they have two sons, Kenneth Alan (deceased) and William Colin (Bill); daughter Gail C. Winborne; granddaughter, April C. Massey; four grandsons: Robert Winborne, Kenneth Alan Clanton Jr., Colin and Blake Clanton (brothers); Great-grandson Kenneth Massey.

He is now retired from ACL RR (16 years) and 29 years with US Postal Service in delivery, customer service and postmaster. He plays golf four days weekly. He has two hole-in-ones.

JAMES CLIFFORD, born Dec. 23, 1918 in Cambridge, MA, entered the USN in March 1944. He attended boot camp in Sampson, NY, followed by special training in Newport, RI where he learned firefighting skills and damage control.

Assigned to the USS *Missouri,* he was on board when the battlewagon was commissioned in Brooklyn, NY. Following this event, the ship departed for Trinidad for a trial test run and training for approximately 30 days. After training, the ship returned to Brooklyn and remained for a lengthy interval before departing for the Pacific.

The most poignant memory of his days in the USN was the Japanese surrender on Sept. 2, 1945 with General MacArthur on board. He feels proud and privileged for having served on the *Missouri*.

He was discharged from the USN in November 1945, having earned three major engagements during time in the service.

JAMES EDWARD CLUCK, born Oct. 26, 1930 in Coffeeville, KS, was inducted into the USN Sept. 30, 1948 and assigned to the USS *Missouri* Jan. 3, 1949, stationed in the fireroom.

His memorable experiences include running aground at Chesapeake Bay. While at sea he also served on the USS *Wisconsin*.

His awards include the Korean Service, United Nations Service and Occupation Award. He was discharged as BT2.

He is married to Toni and they have five children and 12 grandchildren. They are now enjoying retirement in Florida.

JEFFREY DALE COLE, born Nov. 22, 1970 in Memphis, TN, joined the USN June 16, 1989, just 10 hours after his graduation commencement exercises in Portland, OR. He reported to the USS *Missouri* as firecontrolman third class on Aug. 11, 1990 and was assigned to F-3 Div. and later put in charge of Sky-1, 5" gun battery director.

He served aboard *Missouri* from August 1990 to February 1992 which included one six month deployment to the Arabian Gulf in support of Desert Shield and Desert Storm receiving the Combat Action Ribbon and Kuwait Liberation Medal. His most memorable experiences were the Tomahawk launches against Saddam Hussein at the beginning of Desert Storm.

He detached from the USS *Missouri* Feb. 1, 1992 and now lives in Amarillo, TX with his wife Juanita and two sons, Martin and Cordell. He works as transportation manager for Gerald Cole Trucking, Inc.

CHAD H. COLLINS, born Aug. 2, 1969 in San Mateo, CA, joined the USN Oct. 23, 1990 and assigned to USS *Missouri* Feb. 21, 1991, stationed as fireman, E Div. as interior communications electrician.

While at sea he served on board the USS *Missouri* during the Gulf War. He achieved the rank of E-2 fireman apprentice and holds current rank of IC2.

His memorable experiences: just out of boot camp, he found himself in transit aboard the USS *Tripoli*. On February 18, the *Tripoli* struck a mine. His brief stay on the *Tripoli* earned him a Combat Action Ribbon, his first. A day later, he was taken aboard the USS *Missouri*. Not long afterwards, on February 25 the words "brace for impact" came over the *Missouri's* speakers. They had come under direct attack by the Iraqis. The first of two Silkworm missiles had been fired at the *Missouri*. The saddest memory as a USS *Missouri* crewman, was being part of the decommissioning crew. He realized the end of the battleship era had arrived.

His awards include the Combat Action Ribbon, Navy Unit Commendation, Battle E (three awards), Good Conduct, National Defense, Armed Forces Expeditionary Medal, Southeast Asia Ribbon, Armed Forces Service Medal, Sea Service (three awards), NATO Medal, Kuwait Liberation (Saudi), Kuwait Liberation (Kuwait).

He is an enlisted surface warfare specialist

(ESWS). Presently on active duty, USS *Port Royal* (CG-73), Pearl Harbor, HI. Collins is married to Cheryl Cowell and they have one son, Avery James.

VINCENT R. COLLURA JR., born Jan. 27, 1933, joined the USN April 22, 1952 and was assigned to the USS *Missouri* July 1952, stationed in deck/stores.

While at sea he participated in Korea and served on the USS *Missouri* three years.

His memorable experience includes being one of the last 15 men to leave the USS *Missouri* after mothballing in Bremerton, WA.

Collura was discharged April 11, 1956 as SK2. He received the National Defense Service Medal, United Nations Service Medal, Korean Service Medal w/3 stars, China Service Medal (ext.) and the Good Conduct Medal. He is now playing pool to the best of his ability.

WILLIAM H. CONDON, born May 20, 1923, Dorchester County, Eastern Shore, MD, enlisted in USN, January 1942 and served on submarines before boarding the *Missouri* for the commissioning June 1943, where he served until discharge October 1945.

He admired both Admiral Halsey and Capt. Callahan, saying the Admiral was one of the finest and most down to earth persons he had ever known.

Due to some wonderful people in the association he was allowed aboard *Missouri* on Sept. 2, 1995. It was an emotional moment, one he deeply cherished.

Condon passed away Nov. 2, 1995 having spent the day at the Alamo, just two months after his wonderful visit on the *Missouri*.

He leaves his wife, Julia B. Condon, a son, Bill Jr., two daughters, Glenda Kale, Claudia Daniel and seven grandchildren.

RICHARD E. (MOOSE) CONNER, born Dec. 2, 1926 in Streator, IL, entered the USN in 1944. His stations include: boot training, Great Lakes, IL; Radio School, University of Chicago; USS *Pecos*; USS *Missouri*; CR Div.; Iwo Jima; Okinawa; Japan and the surrender.

He is married to Jane Dutler and they have seven children, four boys and three girls. He is now semi-retired. He was in public relations for an attorney. Served on the city if Streator City Council for 28 years, 24 as councilman and four years as mayor.

His most memorable experience has been the privilege of having served aboard the *Mo* and to have had the good fortune to have known so many of the crew, and to have so many great friends, from the organization. Absolutely the greatest people in the world.

DONALD R. CONNOR, born June 9, 1933 in New York, NY, was inducted into the USN April 23, 1952 and assigned to USS *Missouri* as commissaryman, cook/baker/storekeeper. He served in Korea 1953-1953 and condition watches 5"/38.

He was awarded the National Defense, Navy Good Conduct, European Occupation, China Service, Korean Service w/2 Stars, United Nations Medal, South Korean Presidential Unit Citation.

There were many humorous stories about the "galley" and its crew; one recollection, They were under turret #3 when CIC spotted a target, it let go and 50 pound cakes for Christmas sunk; he was making their usual "delicate" 400 lbs. of potato salad when again CIC found another target. Turret #3 let go, the concussion blew a light bulb to pieces over the tub of potato salad. He called Chief Weeks; he called medical, the result, 400 lbs. of "chum" in the sea of Japan.

Practical jokers are always with you; Bill Green, in the bake shop sent one of the watch to the Issue Room (won't mention the name), for "Marbles" for the marble cake. Chief Kalman yelled over "tell Green to cut that — out." Some others were, "The Sea Bat," "The Bucket of Steam", "The Golden River" etc. (Darken Ship), punch cans, threw them overboard on the fantail, it's about 20 degree and 28 knots, a long dark walk, you can't see the fantail, wind; result, over-the-side, berry cans and all (the division "boats" are still looking for them).

Connor is married and has five children and five grandchildren. He retired as sheet metal worker and is now a real estate broker in upper New York State.

HARRY F. COOKE, born March 25, 1927 in Chicago, IL, joined the USN Sept. 25, 1943 and was assigned to USS *Missouri* April 1944, stationed on 4th Div., as deck hand.

While at sea he participated in Iwo Jima Okinawa, and shelling of Japan, just before the A-bomb was dropped.

His memorable experiences: shakedown cruise; squeaking through Panama Canal; the firing power of the USS *Missouri* in action.

He was discharged March 24, 1947 as S1c and received the WWII Victory Medal, Asiatic Campaign w/3 Stars, Navy Occupation and Philippine Liberation Ribbon.

Cooke is married to Shirley and they have seven children: Renee, Terry, Tom, Kevin, Maureen, Dan and Terry. He retired after 20 years in Navy as lieutenant commander. Today he is retired after 42 years of lithography.

GEORGE E. COOPER, born June 7, 1928 in Central Falls, RI, joined the USN January 1946 and was assigned to the USS *Missouri* March 1946, as radio operator.

He was discharged November 1947 as S1(RM) and awarded the Victory Medal.

His memorable experiences: crossing the Arctic Circle on Nov. 30, 1946 and being hit by friendly fire with a five inch shell on the starboard side of the 03 level, resulting in the death of a shipmate. The shell was fired from the cruiser USS *Little Rock;* crossing the equator on Sept. 11, 1947 while returning from Rio de Janeiro, Brazil with President Harry S. Truman on board along with his wife Bess and their daughter Margaret. They all became shellbacks together.

He is now retired from the printing and typesetting industry after 37 years and enjoying every moment of it.

He has been married to Eva V. Marquis since June 25, 1949 and they have one daughter, Linda S. Cooper and one son, George E. Cooper Jr. and two grandsons, Scott and Jeffery.

MICHAEL J. CORRADO, born Dec. 30, 1925, was a member of the 7th Div. and was on the 5.38 inch mount seven gun crew.

He entered the USN in March of 1944 as an apprentice seaman at Sampson Naval Base in New York, was 18 years old. From March 1944 to January 1946 he was on the ship. During those years he had many interesting experiences. He enjoyed himself both with shipmates and officers. Corrado was discharged on Jan. 17, 1946 from Lido Beach, NY.

On shakedown in Trinidad, he had the scare of his life. He was at a beach party with a third of the compliment of the ship. Their menu was "Spam" sandwiches, beer and Yoohoo . . . a real treat! He spotted a coconut tree with a "lovely bunch of coconuts" on it. Corrado pointed it out to his buddies, who challenged him to climb up and get a couple. He climbed up the tree even though it was hard climbing, as he reached the top with his right hand on a coconut . . . a long red tongue flew out at him and almost licked his nose. He was shocked because it was so camouflaged and he slid all the way down the tree like a fireman coming down a firepole. His chest and arms and legs felt the brunt of the slide. His shipmates wondered what had happened. The iguana had made his point.

He is currently retired and plays accordian at the local veterans hospital. He's been married 49 years and has five sons and one daughter. He had his own musical trio that played professionally for about 20 years. Hopes to see his shipmates at one of their get togethers. He misses them.

EDWARD J. CUMMINGS, born Aug. 28, 1920 in Baltimore, MD, entered the USN June 19, 1938 and was assigned to the USS *Missouri* mid-April 1944. Served with FM Div., 3rd Div., as main battery officer, air defense officer, assistant gunnery officer.

His memorable experiences: his forward quarterdeck watch during commissioning; tracking the "bogey" with the MKIA computer which was controlling the guns the night they  shot down their first enemy plane; the kamikaze that hit below Spot II, his main battery battle station; his thoughts the night they were anchored in Sagami Wan before entering Tokyo Bay. As he stood on deck watch-

ing the last glimmer of the setting sun, he realized it was his 25th birthday. He had survived over three years in a World War; a ship sinking in combat; combat wounds; enemy attacks in the Atlantic, Mediterranean and Pacific. He knew many had been through more and many were no longer with them. However, for Cummings, the war was over and he could live in peace. What a wonderful birthday present.

His awards include the Purple Heart, Navy Commendation, Meritorious Service, Bronze Star w/V, and Legion of Merit. He was discharged July 1, 1972 as captain.

Cummings is married to Ethel Crean and they have one daughter, Candace. He is now retired.

FRANCIS R. DALE, born Oct. 30, 1918 in McLeansboro, IL, joined the USN Feb. 19, 1941 and was assigned to the USS *Missouri* Dec. 26, 1947. He served as electricians mate while at sea, firing main battery at Pohang and Inchon, Korea.

His memorable experiences include grounding in Chesapeake Bay, VA. He was EMC in charge of the generating plant.

He retired Nov. 1, 1962 as EMC and received a commendation for maintaining electrical power during grounding.

Dale has been retired since Jan. 2, 1981 from the San Diego County Sheriff's Department as bailiff in superior courts. He has been married to his wife for 54 years and they have one daughter and five sons.

W. KENT DAVIS, born Dec. 22, 1962, in Montgomery, AL, graduated from LSU in December 1985, joined the USN Jan. 31, 1986. After training at OCS and Supply Corps School, reported aboard USS *Missouri* (BB-63) in March 1987, where he served as disbursing officer, ship's store officer, and 5" gun director officer. While aboard *Missouri*, he married Lisa Rogne. He was stationed on the *Missouri* until December 1989, when he transferred to NAS Atlanta as assistant supply officer. Left NAS Atlanta in March 1992 and reported aboard USS *Abraham Lincoln* (CVN-72), serving as public affairs officer until discharged July 31, 1994, in Alameda, CA as a lieutenant.

He still serves to this day as a lieutenant commander, USNR, in the public affairs unit NR NAVINFO SE 108 in Atlanta. He is also attending law school at Georgia State University. He and Lisa reside in the Atlanta area.

WALTER F. DEITZ, born Dec. 1, 1929 in Camden, NJ, was inducted into the service Aug. 2, 1948 and assigned to the USS *Missouri,* March 15, 1949, #3 Distribution, #3 Engine Room.

He received the United Nations Service Ribbon, Good Conduct Medal, Korean Service Ribbon, Navy Occupation Ribbon, Republic of Korea Presidential Unit Citation and the National Defense Service Medal. Deitz received his discharge June 9, 1952.

His memorable experiences include: April 1949 the USS *Missouri* took a reserve cruise to Panama. He

was very familiar with the sun and sunburn from going to the New Jersey shore every summer of his life. One day after noon chow he went up on the fantail to get some sun and disaster struck. There is a big difference between the New Jersey sun and the Panama sun. He was a mess, but he really didn't realize just how bad it was until the next morning when he tried to get dressed.

He had just come aboard a few weeks before with what they called the "D" draft. There were ten of them with the last name beginning with D that came fresh out of electrician's mate school in Great Lakes and they were assigned to live in the overflow compartment way back aft. Those guys were all great to him, helping him to get dressed and reminding him that if he went to sick bay he would be written up. What a mess!

His work station was #3 distribution in #3 engine room and he was working for a third class by the name of Davis who was also very kind to him and did not make him work, but finally after a couple of days he had to pull watch in #4 engine room on the generator switchboard. It was a 4 to 8 and the switchboard in the engine room is raised about two feet higher than every thing else and he was very visible. Perhaps it was the heat or the pain or whatever but he was throwing up pretty often in the GI can for most of the watch. As he stated, he was pretty visible and soon the chief machinist mate in charge of the watch saw him and came over. He was finally caught.

"What's the matter kid, if you're sick go to sick bay," the chief told him.

What else could he do but tell him the truth and see what would happen. Deitz will never forget that guy. He can't remember his name and he stood many watches with him. The whole watch he would walk round and round the engine room and seldom stop.

"OK kid when you finally get off the watch here's what you do, go up to the galley and get a gallon jug and fill it with water. Then go to the steaming fire room and sit behind the boiler in the upper level as long as you can stand it. Get someone to keep filling the water bottle and keep drinking".

Now, in the Navy less than a year, on the ship less than a month, was this chief playing a joke on him? Was this some kind of initiation? Was he nuts enough to try this? Yes, he was hurting so bad and so sick nothing was too nuts to try.

The cooks in the galley looked at him a little funny, wondering what this nut was going to do, but they gave him the gallon jug. He filled it with water and went to the fireroom and the boiler tender looked at him kind of funny but they had done some electrical testing in this fireroom and he knew Deitz a little so he didn't chase him out, especially when he saw his stomach and chest. Shipmates will help you out of trouble no matter how nuts they think you are.

If you have a bad sunburn and can work up a sweat, you'll look terrible, but all the pain will be gone. He could actually see the blisters fill up and continually break. The guys on the watch kept coming back and checking on him and refilling the gallon jug. He lasted almost a half hour and when he came out of there it looked like he had leprosy, but all the pain was gone. After a shower and a good nights sleep he was able to dress himself and tie his shoes. He was one happy sailor. You can bet he found that chief and thanked him first.

He married Joyce S. Stockton Oct. 6, 1951 and they have two children and two grandchildren. Deitz retired from Philadelphia Naval Shipyard Jan. 3, 1985 with 32 years of service.

JOHN J. DELANEY, born Sept. 5, 1934 in Troy, NY, enlisted in the USN April 1952 and served in USS *Missouri* 1952 and 1953. During his USN career he saw service in Korea and Vietnam. Other duty stations include: USNAS Jacksonville, FL; USNH Yokosuka, Japan; Naval Advisory Group, Korea; Naval Ammunition Depot, Earle, NJ; USS *Oriskany* (CVA-34); USS

Lexington (CVA-16); instructor, USNRTC Binghamton, NY; USNS Geiger TAP-197; USS *Barry* (DD-933); 3rd Marine Div. Vietnam and Cruiser Destroyer Force Pacific. Retired as chief hospital corpsman October 1971.

During his career he earned the Navy Achievement Medal w/Combat V, Combat Action Ribbon, Navy Unit Citation, Navy Good Conduct w/5 Stars, Navy Expeditionary Medal, China Service Medal, National Defense Medal w/Star, Korean Service Medal w/2 Stars, Vietnam Campaign w/4 Stars, United Nations Service Medal, Vietnam Service Medal, Vietnam Gallantry Cross w/Palm and Frame, Korean Presidential Unit Citation, Vietnam Presidential Unit Citation and the New York State Conspicuous Service Cross.

Delaney is married to Marjorie Dauntler and has two children, Timothy Delaney and Marion Delaney-Driscoll, and two grandchildren, Alyssa and Ashley Driscoll.

After retiring from the USN, he was director of support services at a Troy, NY hospital. He retired in 1987 and moved to Minden, NE.

EDWARD ORLANDO DE PHILIPPO SR., born Feb. 23, 1934 in Bangor, ME, joined the USN July 30, 1951 and was assigned to the USS *Missouri* October 1951, stationed with 1st Div., left gun layer 16".

While at sea he participated in the Korean conflict. His memorable experiences: in a 5" gun mount off the coast of Korea and they were at GQ. The sea was rough and their speaker was not working in their mount. All hands were told to stay off the main deck, but he and a buddy, David A. Bradley went to get some hot soup below in the mess hall. That was a big mistake. As they opened the hatch and got out on deck, a wave of water about 30 ft. high hit them and they took a ride about 75 to 100 feet. He thought they were gone. His buddy just happened to reach out as he grabbed De Philippo, and caught some line or hatch and when that wave stopped, they ran below before the next one hit. If David had not have grabbed him they would not be here today. Thanks David and thank God.

He was discharged Dec. 29, 1954 as SA and received the National Defense Medal, Navy Occupation, Korean Service and China Service Medal.

He is married to Faye C. Chapman and they have three sons: Ed Jr., Kirk and Mark. De Philippo is disabled because of cancer, but has been in remission for 16 years. He continues to get shots every month.

MANUEL R. DESOUZA JR., born Feb. 25, 1924 in Norton, MA, reported aboard the *Missouri* before Christmas on delayed orders. He lived aboard in New York. The paymaster changed that date later. His assignment, B Div. mostly #4 fireroom. He was also assigned to A Div., all auxiliary units, operational watches engine rooms and main engine con-

trol - boiler makers repair rolled tubes and brickwork; boiler maintenance, security throughout all spaces "OD" watches as required. Many shore patrols.

His memorable experiences include: tour guide many times, (VIP and general visitors); interpreter; called to the bridge, pilot didn't speak English. Pilot was taken sight on land markers prior to anchor. He was not ready. But CO said, "Let go anchor". Then later they had to move, bearings were off. Ship needed more swing room to clear other ships. The river runs both in and out at the same time.

He was sent for, and reported to CO cabin. A special guest, French Bishop and his people. Three men helped him get ready in 10 minutes. Desouza talked Portuguese-Spanish, some Italian and got over whatever to get the job done.

He was interviewed one time in an office. They asked him questions to interpret. He gave his best and when it was over, the man was US Representative Embassy and Expert. He turned to Chief of Police, also Army General and said, "Good enough."

DeSouza was awarded 11 Battle Stars, Unit Citation, Good Conduct, National Defense, China Service, Asiatic-Pacific, Aleutians and Alaska, Korea, United Nations, Philippine Liberation and the American Victory.

He was discharged from the USN in May 1960, has been married 43+ years and has two daughters, three sons, and 10 grandchildren. He is now retired.

IVAN D. DEXTER, born Aug. 26, 1925 in White Plains, NY, enlisted USN Aug. 19, 1943. His stations include: boot camp, Newport, RI; Hospital Corps School; Portsmouth VA Hospital; LST 138; SS *Thomas Johnson;* Philadelphia, PA; USS *Missouri* September 1944 to March 12, 1946; PSC; Lido Beach, Long Island, NY. Total sea duty, 29 1/2 months.

He was awarded the American Theatre Medal, Asiatic Pacific w/3 Stars, European w/Star, Occupation Medal and WWII Victory.

Action at sea: Europe June 6, 1944, D-Day; Utah; Omaha Beach; Cherbourg, France; served on the *Missouri* in Okinawa, Iwo Jima, B/B of Japan. He was part of regimental staff landing force at Yokosuka Naval Base just inside Tokyo Bay, during and after the surrender.

His most memorable experience occurred on the way home. After Japan's surrender, they stopped at Hawaii. They all stood at the ships rail in their dress whites. The people went nuts over them. As the ship slowly moved to the dock the band played the Hawaiian war chant. It was a very emotional time for him. He got goose bumps all over. Thank God the war was over and they were on their way home. He thought maybe now they could have peace on earth and good will to men.

He married Anne Ferschke and they have two daughters, Dianne and Deborah and one grandson, John. Dexter is now retired from the Firearms Fund Insurance Company, he was the inland marine manager of the New York office.

ARTHUR G. DIXON, born Oct. 7, 1925 in Wells, ME, enlisted in the USN Aug. 18, 1943. His stations include: boot camp and Gunner's Mate School, Newport, RI; patrol group for submarines from base at Southport, NC; Charlestown, NC Navy base; USS *Missouri,* 8th Div., Gunner's Mate Section.

Being a gunner on a 20mm gun mount port side of turret #3 gave him a front seat to all action. Highlights to him were the USS *Franklin* going up in

flames from a Japanese bomb; taking a kamikaze hit to their own ship; and being a witness to the unconditional surrender of Japan on the deck of the *Missouri.*

He received the American Theatre, Asiatic-Pacific w/3 Stars, Victory, Occupational Service WWII Victory, Philippine Liberation, Philippine Independence, and Good Conduct Medals.

Dixon received an honorable discharge, Oct. 24, 1945 from USN Personnel Separation Center, Boston, MA.

He has been married to Edith Harris Dixon for 52 years and they have one son and daughter. He retired as a line supervisor from Central Main Power Company in 1987 after 40 years of service.

WILLIAM A. DIXON, born Dec. 11, 1926 in Nutter Fort, WV, enlisted in USN in December 1943 at age 17. Served on Destroyer 493, on Minesweeper 33 and battleship *Missouri* (BB-63).

Entitled to the following awards: Navy Good Conduct, American Campaign, Asiatic-Pacific w/ Battle Star, EAME w/2 stars, WWII Victory Medal, Navy Occupation Medal w/Asia Clasp, National Defense, Korean Service Medal w/3 Stars, United Nations Service Medal, China Service, Korean Presidential Citation, Navy E Ribbon and Battle of Atlantic from Great Britain.

He served aboard the USS *Missouri* from April 1948 to April 1952 and was main battery range finder operator at sea detail, condition watch and general quarters spot 1. Was FCI when discharged. Retired as civilian employee from Army Corps of Engineers.

ERNEST L. DOREY, born March 14, 1927 in Burlington, VT, was inducted into the USN March 7, 1945 and began active duty May 9, 1945. He was assigned to USS *Missouri* March 22, 1946 in the machine shop. His stations include Sampson, NY; Camp Bradford; Little Creek Annex 3, VA; Melville, RI; USS *Missouri.*

He was discharged Aug. 2, 1946 in Boston, MA and awarded the WWII Victory Medal and American Area Medal.

Dorey is married to Lorraine Richards and they have two sons. He retired from General Electric Company May 29, 1986 after 37 1/2 years.

ROBERT S. DORKO, born Dec. 26, 1931 in Binghampton, NY, joined the USN March 26, 1952 and assigned to USS *Missouri* June 1952, serving as firecontrol man, FA Div.

While at sea he served a tour of duty during the

Korean War. He was discharged as FT3 March 23, 1956.

His memorable experiences include the sadness of losing three shipmates during their Korean operations, also the passing away of their skipper. Happier times include the midshipmen cruise to Rio de Janeiro; crossing of the equator; passage of the Panama Canal; the great welcome they received in San Francisco and Seattle while enroute to Bremington, WA to decommission their ship; the pride and devotion to duty displayed by his fine shipmates while serving aboard the *Mighty Mo.*

Dorko received the Korean Service Medal, United Nations Service Medal, China Service Medal and the Good Conduct Medal.

He is married to Connie and they have three children: Robert Jr., Cynthia, and Michael and five grandchildren: Stephine, and Emily Dorko; Jennifer, Cristin and Matthew Bronson. He is now retired from IBM Corp, Endicott, NY after 26 years of service and resides in Johnson City with his wife Connie.

JOSEPH G. DOWNEY, born Jan. 4, 1927 in Boston, MA, joined the USN July 2, 1942. Served on the USS *Texas* 1942-1943 and assigned to the USS *Missouri* April 1944. He served with 11th Div. 40mm AA and achieved the rank of S1c.

While at sea he saw the USS *Franklin* hit and two kamikaze hit the *Missouri.* He was member of 3rd Fleet Landing Force.

His memorable experiences include commissioning of the *Mo* 1944; Harry Truman viewing fleet in 1945.

He was discharged April 21, 1947 and received the Good Conduct Medal, American Theatre, Victory Medal, EAME w/star, Asiatic-Pacific w/3 stars and Occupational Medal w/clasp.

Downey now lives in Florida and is self-employed in a nautical gift shop.

JOHN A. DUBENSKY, born Aug. 16, 1922 in Delancey, PA, moved to Ambridge, PA where he grew up and graduated from Ambridge High School.

He enlisted in the Marines in November 1942 and attended boot camp at Parris Island. Stationed in Washington, DC; Sea School Norfolk, VA. Assigned to USS *Missouri* (Brooklyn Navy Yard) was on board as Margaret Truman commissioned the *Missouri.* He was discharged November 1945.

His duties include captains orderly, 20mm gunner during GQ. Achieved the rank of corporal. While at sea, attacked by Japanese planes in the Pacific on numerous occasions.

His memorable experiences include the kamikaze (suicide plane) hitting the *Missouri;* surrender of Japanese and signing of the peace treaty Sept. 2, 1945.

He attended re-commissioning in San Francisco, and de-commissioning, also the replacement of the surrender plaque at Long Beach.

He married Mary Karaffa. Retired from Datatape (company of Kodak) as supervisor of the tool & die makers.

STEVEN MARK DUDLEY, born Sept. 28, 1966, grew up in Palo Alto, CA and entered the Naval Academy on July 2, 1985 and graduated in 1989, selecting Surface Warfare for his career path, and the *Mighty Mo* for his first ship assignment. He was initially assigned to the 4th Div., responsible for main deck and 01 Level areas, including the surrender deck, as well as several boats.

After *Missouri* returned from the Gulf War, he was assigned to the Operations Department as electronics materials officer, during which time he was responsible for the preliminary removal of electronic equipment, in preparation for the *Mo's* upcoming decommissioning. He has good memories of the exceptional men who served under him. He resigned on Sept. 3, 1993 with the rank of lieutenant.

Memorable experience: times spent sitting out on the top of the 16-inch gun directors, particularly "Spot One," the highest manned station on the ship, during much of the real action of the Gulf War. Since the big guns were being controlled generally by CIC and Combat Plot, there was not much for him and his director crew (Fire-Controlman Mike Anderson and Dan Shock) to do except stay vigilant. Therefore, they would often open up the top hatch of the director and sit out on top of it, as it was large enough for all three of them to sit fairly comfortably. This was a particularly advantageous position, not merely for watching the shore line for enemy tanks and artillery, but especially for getting pictures of all the action that was taking place. They watched almost 15 of the 28 Tomahawk missiles being launched, blasting off just tens of yards from where they were perched. They even saw the one that for some reason did not release from its armored box launcher and just torched the whole deck behind it as its booster rocket depleted itself.

When it was time to get in close and use the big guns, like the time the Iraqis overran the Marines at Ras-al-Khafji, the excitement and the incredible scenes increased. They were able to watch the 16-inch-ers do their mighty work from a vantage point matched not even by the bridge (for they also had a fair view of turret 3's actions, and they didn't get all the smoke from the blasts). They didn't get to see the results, except on the Site TV system, though, because the targets were pretty far inland, and they were using *Missouri's* remotely piloted vehicle to spot the fire missions. But they saw much else anyway, such as a nighttime firefight a little ways inland from the beach, with tracer rounds lighting the night and the slit-taped lights of HMMV's racing to and from the skirmish.

On moonlit nights they could see the silhouettes of their planes going overhead, either going to or returning from their missions. One night they watched from afar an aerial bombing of what appeared to be some type of industrial installation close to the beach. The installation's searchlights lit up the sky, and they could see the puffs of smoke in the air from the anti-aircraft guns, just like in a WWII movie. It was strange to watch, knowing it was real, but being able to just sit and watch in some detached way, through the director's high-powered optics.

All in all, the *Mo* fired almost 900, 16-inch projectiles, and 28 Tomahawk missiles, and his director crew and Dudley got to see a large percentage of those munitions sent on their way. It was very sobering to think of what was happening to the people on the receiving end, and he remembers just thanking Heaven that the *Mighty Mo* was on their side.

He is now married with one little boy, eight weeks old and is pursuing a career in law enforcement.

THEODORE H. DUNHAM, born Aug. 21, 1931 in Brooklyn, NY, inducted into the USN May 18, 1951, was assigned to the *Missouri* September 1951, ICFC elect. shop.

His memorable experiences include the times he spent with his shipmates.

While at sea he participated in Korea.

Dunham was awarded the Good Conduct, China Service, Korean Service, United Nations and the National Defense. He was discharged as EM1, May 12, 1955.

He is married to Genevieve T. Dunham and they have four children. He is now a retired teacher.

WILLARD FRANK DUNLAP, born Jan. 3, 1926, is a college graduate. He joined the USN March 10, 1944, plankowner USS *Missouri* April 1944, Newport, RI. Precommissioning. Left *Missouri* May 1946. Positions: sun lookout on 130' mast, watching sun for kamikazes. He was highest man on ship not in rank, but in position.

They took part in the Philippines, Iwo Jima, Okinawa and bombardment of northern Japan.

He went on to Tokyo Bay for the surrender ceremony. Very impressive especially with hundreds of planes flying overhead during signing. He went back to Guam, Panama Canal, Norfolk and New York where ship was on display to thousands.

Went to Cuba for replacement training. On to Turkey with the Turkish ambassador body. Stopped at Gibraltar, went through the "Rock".

Departed for Turkey, royal welcome. On to Piraeus and Athens Greece, from there to Naples Rome Italy. Departed for Algiers and Tangiers North Africa. Then home for discharge, a wonderful experience. He has been married for 39 years to Shirley Bell. Retired and world travelers.

BUFORD H. EDWARDS, born in Lebanon, TN, joined the USN April 28, 1944 and assigned to the USS *Missouri* June 10, 1944 and stationed as lookout. He was discharged Dec. 26, 1945 as S2.

While at sea he participated in Iwo Jima, Okinawa and Japan. He is now retired and widowed.

ROBERT G. EICHENLAUB, born Nov. 30, 1929 in Cincinnati, OH, was inducted into the USN Sept. 30, 1948 and assigned to the *Missouri* Jan. 10, 1949, stationed on main battery plotting room.

While at sea he had the opportunity to visit many

foreign countries. Participated in shore bombardment, first Korean deployment.

He was awarded the Good Conduct Medal, Occupation Service, National Defense Service, Korean Service, Korean Presidential Unit Citation, and United Nations Service. He was discharged Aug. 28, 1952 as FT3

Eichenlaub retired from Proctor and Gamble Co.

ROBERT WAITE ELLIOTT JR., born in Ashtabula, OH Dec. 4, 1921, received a Bachelor of Metallurgical Engineering Degree from the Ohio State University, 1943. He worked in the steel industry until October 1944. Joined the USN as seaman recruit. After boot camp assigned to the Naval Reserve Midshipman School, Columbia University. Upon graduation, commissioned ENS D (L). After WWII enrolled in dental school. Received DDS June 1950, Western Reserve Unit. Commissioned Dental Corps. Served in numerous billets ashore and afloat. Last duty, chief of the Dental Corps in the grade of rear admiral. Retired from the service in 1977. Awarded the Legion of Merit.

In October 1952 assigned USS *Missouri* (BB-63) as a junior dental officer, LTjg. Made division officer of DH Div. Left *Missouri* in Norfolk, VA summer of 1954 following midshipman cruise to Europe.

After retirement from the Navy, appointed associate professor at Georgetown University where he taught for seven years. Retired 1984 as clinical professor. Honored by Georgetown with Dr. Sc. (Honoris Causa). Employed in private practice for the next three years. Elliott has lived in Key West since 1987. He married Carolyn J. Blaser in 1944 and they have two children and five grandchildren.

JOHN W. ERPELDING, LT, USNR, FM Div., April 1945-May 1946, born Oct. 17, 1919, Chicago, IL. Enlisted in Navy V-7 Program Jan. 4, 1942 while a student at University of Michigan at Ann Arbor, MI. Had tried before Pearl Harbor to enlist in Navy Air Corps. Passed all tests except physical (they said I had flat feet); I protested to no avail stating that I was on the Michigan Varsity football and baseball teams.

Graduated from Midshipman School at Northwestern University, Chicago, IL, and received commission as an ensign in July 1943. Was retained on staff and taught Ordnance and Gunnery at the Midshipman School until ordered to proceed and report to the *Missouri*. Transferred from a Destroyer to *Missouri* by breeches buoy while both ships underway and reported aboard in April 1945 as a LTjg and was immediately assigned to the quad 40mm gunmount and director on the fantail known as "Kamikaze Corner."

Shortly thereafter, was assigned to the FM Div. Main Battery Fire Control as the fire control officer in Spot 3, Battle Station in the Fire Control Tower with the gunnery officer, Cdr. Bird. Volunteered for the 3rd Fleet Naval Landing Force and was assigned by Missouri Executive Commander Malone (the commanding officer of the 3rd Fleet Naval Landing Force) to replace another officer who was ill. He was assigned to the Battalion HQ staff and landed August 30 at the Yokosuka Naval Base in Tokyo Bay. They were met on the beach by a Japanese Lt. Cmdr. who bowed to the American officers. They raised the American flag on the Japanese naval base by formal ceremony Aug. 31, 1945. He was able to get back to the *Missouri* for the Sept. 2, 1945, formal surrender

ceremonies. He was officer of the deck on Navy Day October 1945 in New York when President Truman, Governor Dewey of New York and Mayor LaGuardia of New York City came aboard.

He was privileged to be aboard the *Missouri's* "good will tour" (saber rattling) from March-May1946 where the ports of call were Gibraltar, Istanbul, Turkey; Athens, Greece; Naples, Italy; Algiers and Tangier. Departed *Missouri* June 1, 1946 and was discharged July 8, 1946 as full lieutenant from Great Lakes NTS.

Entered Northwestern University Law School in Chicago, IL on GI Bill and graduated in August 1948. He has practiced law in the state of California continuously since 1950. John has been a widower since December 1993. He has been blessed with four lovely and loving daughters and grandchildren.

Benediction: The magnificent *Missouri* has her place in history and those of us who ever served aboard her share a part of that history. To me it is a great honor to have been a member of the *Missouri's* Ship's Company, i.e., the officers and crew of the *Missouri*, an elite group of men whose duty performances, individually and collectively, rate that special Navy commendation for superior performance "well done." I salute you each and everyone.

HERBERT FAHR JR., born March 14, 1933 in Brooklyn, NY, enlisted in the USN June 4, 1952. Trained at Bainbridge, MD, stationed at Newport, RI Underwater Ordinance Station, Gould Island, Submarine School at New London, USS *Missouri* (BB-63), USS *Eldorado* (AGC-11) and USS *Washburn* (AKA-108). He received the National Service Medal, European Occupation Medal, China Service Medal and the Good Conduct Medal. He was honorably discharged on June 3, 1960.

Fahr went to school on the GI Bill in Milwaukee and then worked in Japan for three years. Upon his return he joined a manufacturing company in their engineering department and finished college at night graduating from NY Institute of Technology with a BS in engineering.

He has served as newsletter editor, vice president and president of the USS Missouri Association. He enjoys traveling and is a computer nerd. He and his wife Arlene are retired and enjoying the good life.

MATTHEW J. FERGUSON, born Feb. 22, 1927, New York City, NY, enlisted in the USNR on Jan. 23, 1945, and went to boot camp in NTC Sampson, NY. Upon completion of training in six weeks, requested sea duty, his three choices and was put out to sea on the USS *Missouri* until discharged July 1, 1946 at PSC Lido Beach, LI NY.

Served as fireman second class and then first class in E Div. in engine room 3 and the aft diesel room. They could hear sea battle accounts over the loud speakers during the bombardments at Okinawa and Honshu Japan.

Ferguson received the American Theatre, Asiatic-Pacific w/2 stars, and the Victory Medals. He also received the EM3c rating during the Mediterranean

cruise to Turkey after the war. Saw Pope Pius XII, and came home.

He married Patricia A. Thompson, Oct. 15, 1950 while attending Columbia University in New York. They have two sons. One graduated West Point in 1980, the other son is an AFL/CIO driver of heavy equipment. Ferguson retired from Grumman Aerospace Corp. after 27 years on Feb. 22, 1990. Later worked for the Veterans Administration in engineering. He still maintains licenses as a professional engineer in New York and Indiana and still loves sailing his Catalina 27. He was on Long Island Sound July 1997, 51 years after leaving the *Big Mo*.

WILLIAM FERKAN SR., born March 18, 1926 in New Bethlehem, PA, joined the USN March 17, 1944.

He was awarded the Pacific Theatre Ribbon w/3 stars, American Theatre Ribbon and the Victory Medal.

Ferkan died March 8, 1995 of a massive heart attack after snow plowing two neighbors and his own drive ways. He is survived by his wife Lois and two sons, William Jr., and John.

WILLIAM (BILL) FIGURA, born June 27, 1926, Binghampton, NY. Joined the USN Feb. 17, 1944; boot training, Sampson, NY; Newport, RI; assigned to the USS Missouri June 11, 1944 as gunner's mate.

Participated in Iwo Jima engagement, shelled Okinawa, Kyushu. Memorable experience was kamikaze attack above his battle station.

Discharged May 13, 1946 as S1/c. Received American, Asian and Victory Medals w/3 stars.

Became a baker with apprenticeship at Schmidtz Bakery and eventually opened his own Figura Bakery. Married Anne in 1948 and has four boys, five girls and 13 grandchildren. Bill passed away Sept. 20, 1982.

JOHN THOMAS FINN, born Sept. 23, 1925, of East Farmingdale, founded the USS Missouri Association in 1974 with a crew of 60 servicemen and their wives. Today, 24 years later with membership of over 1,200 to commemorate the battleship on which the Japanese surrendered to end WWII.

He enlisted in the USN on his 17th birthday in 1942 and was stationed at Newport, RI, first serving aboard the USS *Nevada*, until the USS *Missouri* was commissioned. He served on the *Missouri* until his discharge on Nov. 24, 1946.

Many memorable moments to remember, but uppermost was when the Japanese suicide plane came barreling into the machine guns. Best memory of all was standing on the deck, Sept. 2, 1945, witnessing the signing of the surrender of the Japanese to the US and ending WWII. HE knew it was a historical event that would be remembered for many years. This was why he started the USS Missouri Association, to continue the camaraderie of all his shipmates and future shipmates who served aboard the USS *Missouri*.

He married Dorothy and they had a son, John T. Jr., daughter, Katherine and five grandchildren. Finn died of a cardiac arrest April 14, 1987.

HERBERT HAYDEN FLAGG, born Dec. 2, 1926 in Schenectady, NY, joined the USN in March 17, 1944 and was a seaman first class C1, assigned to the USS *Missouri* June 18, 1944.

He saw action at Okinawa, Bonin Islands, Tokyo Bay, Guam and Leyte. His most memorable experience was pulling into Tokyo Bay on the *Big Mo*.

Flagg was discharged May 13, 1946 and received the American Theatre Medal, three Pacific Stars and the Victory Medal.

He worked in the Scotia Glenville Schools and retired in February 1994. He passed away on Aug. 10, 1996 and is survived by his wife, Ann, sister Joyce, two sons, Robert and Stephen, two daughters, Cynthia and Teresa and five grandchildren: Renee, William, Daniel, Shawn and Daniel.

In September of 1995 he and his wife flew out to Seattle and it was one of the most exciting experiences of her life to go aboard the USS *Missouri*.

DONALD M. FLAHERTY, born Dec. 14, 1928 in Janesville, WI, joined the USN June 1945 and was assigned to USS *Missouri* from January 1948 to February 1951. He served in R Div. (Damage Control Shop) and achieved the rank of DC2c.

While at sea he served in Korea. Transferred to USS *Halsey Powell* where he was wounded and received a medical discharge.

His memorable experiences include being stuck in the mud in Hampton Bay; going through Panama Canal to Korea.

He was discharged October 1951 from the Great Lakes Naval Hospital and received the usual BAR ribbons, Good Conduct Medal, Purple Heart and Bronze Star.

Flaherty is married to Pat and they have sons, Dave and Mike, a daughter, Carrie and two grandchildren. He is retired after 25 years with Department of Transportation with the State of Wisconsin.

PAUL A. FLAMM, born Feb. 2, 1927 in Reading, PA, joined USN Feb. 1, 1945 and was assigned to USS *Missouri* June 1945, as head electrician 1# 16" turret.

While at sea he participated in the bombardment of Okinawa and islands of Japan.

He was discharged July 1, 1946, as petty officer third class.

Flamm received the Asiatic-Pacific area and Victory Medal w/2 Stars.

His memorable experiences include his interest in all the electrical equipment in his charge; swinging like a pendulum while climbing a ladder in the #1 "16" turret as the result of a kamikaze which

exploded off of their fantail; crossing USA via troop train from Sampson, NY to San Francisco area; cruise to Istanbul, Greece, Italy, etc.

He is married to Barbara and they have a daughter, Caroline (grandchildren Drew and Benjamin Bastian); son Eric (grandchildren Sean and Mandy Flamm). Flamm does volunteer work in his community. He is a world traveler, enjoys playing golf and retired in 1986.

He attended Reading High School, Penn State University, BS; Temple University, M.ED, Lehigh University, Superintendents Letter, Nova University, FL, Ed.D.

Flamm served as a teacher, principal, superintendent (Oley Valley School District, Oley, PA 12 years). Thirty-five years in education.

JOSEPH C. FLORY, born April 11, 1926 on a farm near Manheim, PA, joined the USN in March 1944, three weeks before his 18th birthday. Took boot camp at Sampson, NY, *Missouri* training at Newport, RI, and assigned in the ship on June 11, as one of the original crew. At sea he was assigned on a 20mm mount on the fantail.

Things he remembers while at sea were being hit by a kamikaze, participating in three major battles; sailing into Tokyo Bay and witnessing the signing of the Japanese surrender.

After the surrender he was transferred to the SS Div. and they made the Mediterranean mission to Turkey, Greece, Rome, Tangier and Spain.

Flory was discharged on May 12, 1946 at Bainbridge, MD. He married a shipmates sister, Anna L. Wetzel and has two children, six grandchildren and three great-grandchildren. He was employed as a hot press operator at R.M. Friction Company of Manheim. Retired July 1988 after 41 years of service. Enjoys gardening, auctions and bus trips.

He belongs to the USS Missouri Association and has attended most of the reunions since 1979. Flory is a member of the Navy Club USA Ship 91.

THOMAS F. FLUCK, rejected at 16 1/2 by USN enlistment station, he returned days before his 17th birthday in February 1944. After examinations, etc. he received notice to report for active duty a week after his birthday.

He attended boot camp at Sampson, NY. After boot leave was assigned to the "*Missouri* detail" where he attended Signal School and Fire Fighting School at Newport, RI.

He reported aboard the USS *Missouri* prior to her commissioning. After the ship's maiden voyage and shakedown he made seaman first.

Between the Iwo Jima and Okinawa invasions he took exams for signalman third class and passed, being deprived of same for petty "chicken reasons".

Disembarked *Mighty Mo* after her trip to Turkey in 1946. Discharged at Lido Beach with three combat Bronze Stars. Received high school diploma the following year.

JACK M. FORD, born Oct. 1, 1927 in Coconut Grove, FL, enlisted in the US Marines Feb. 22, 1943 (note only 15 years and four months old at the time). Not a very uncommon occurrence in the early years of WWII. Education at the time, five months into the 10th grade. Completed boot camp at Parris Island, SC and then assigned to Camp Lejeune, NC for advanced combat training. Stationed at Lejeune until assigned to Sea School in Portsmouth, VA. Completed school with the Marine Detachment USS *Missouri* June 1944. They moved aboard ship a few days before the commissioning ceremony on June 11, 1944. Served aboard *Missouri* until May 1945.

Transferred to the 6th Div. 22 Marines on Okinawa. After the war he was stationed at the Marine Barracks Washington, DC and Daytona Beach NAS until discharged on December 1945. He stayed active in the Marine Corps Reserve until 1950.

Completed education at the University of Miami and retired as supervising engineer P.E. RCDD from Southern Bell Telephone Co. after 37 years. Continued to work as a consulting engineer in Miami, FL for six years prior to retiring to North Carolina in 1988.

He is married to Vivian Farmer and they have one son, Wayne Ford, three daughters: Ruth Cable, Donna Fales, Sue McWhorter and 12 wonderful grandchildren.

A major disappointment was not staying aboard the USS *Missouri* until wars end and missing the surrender ceremonies in Tokyo Bay.

ROBERT FORSETH, the USS *Missouri* has played a big part in his life every day since the day in July 1946 when he boarded her, while she was on port of call and anchored in Casco Bay, Portland, ME.

He was immediately assigned to the M Div., engine room #2. His immediate reaction was how awesome a vessel and how astounding the machinery and manpower needed to operate and maneuver this magnificent ship.

During his tour of duty they made many maneuvering cruises and ports of call. One such cruise was to Rio de Janeiro to pick up and transport back to the United States, President Truman, his family and entourage.

Throughout their cruise, many of the ships company became pollywogs, shellbacks and bluenoses.

In the fall of 1947 the *Big Mo* was on port of call in New York City prior to overhaul in the Brooklyn Navy Yard. About this time he met Virginia, a magnificent young lad who would one day become his wife. They dated until February 1948 when he was discharged and he returned to his home town. They continued to correspond.

In December 1950 he was laid off for the season and returned to New York where he and Virginia (Ginny) had renewed their friendship and in June 1951 they were married.

They raised five children and have been married 46 years.

His interest in ships and machinery has never

abated and upon his retirement he held the rank of licensed chief marine engineer.

He states, "I have to say Thanks *Big Mo*" for you have been a continuous on going part of my life. "We Remember."

CHARLES E. FRASER, born March 10, 1922 in Detroit, MI, graduated in 1940 from Rogers City High School, Rogers City, MI where he played football and baseball. Enlisted in the USN in August 1942 and received basic naval training at Grosse Isle NAS, Grosse Isle, MI. Assigned to the USS *Missouri* and served from December 1944 to January 1946. Member of the 2nd Div. with a rating of second class boatswain mate. His general quarters station was in charge of No. 2 turret powder handling room.

He received the Victory Medal, American Theatre, Asiatic-Pacific w/three stars, Occupational and Good Conduct Medals. Witnessed the signing of the Peace Treaty on board the USS *Missouri* Sept. 2, 1945. Discharged January 1946.

He married in 1947 to Geraldine Frantz and is the father of two daughters and two grandchildren. Fraser retired from construction business in 1984 and has resided in Tucson, AZ since 1959.

SCHUYLER (SKY) FREDRIKSON, born Feb. 11, 1927 in Brooklyn, NY, raised on a farm in Holden, MA, enlisted in USN Feb. 11, 1944 at 17 years of age. Went to boot camp at Sampson, NY, then on to further training at Newport, RI. Boarded USS *Chilton*. From there he went to Brooklyn Navy Yard and boarded USS *Missouri*. He remained on the USS *Missouri* until discharged in Boston, MA on May 13, 1946.

His first duties were as a lookout in L Div. Later, he transferred to R Div. and damage control. While at his battle station on the fantail, he was the first firefighter on the scene when the only kamikaze to ever hit the deck of the USS *Missouri* crashed during an air attack. He cut out the red circle insignia from the wing of the plane with AV shears and later passed out the pieces to his fellow crew mates.

He was selected for the USS *Missouri* diving team along with Verlan Dalton and Jim Lynch. The first dives, to check the four screws and two rudders to make sure no lines or cables were entangled, were made in the Ulithi Islands, better known as "Mog Mog". Although deep-sea hard hat diving gear was available, it was never used. Instead, used diving gear, state of the art for its time, that consisted of a homemade rigging called an RBA (rescue breathing apparatus), bathing suit, lead-weighted belt, and an air compressor on deck. The man topside on compressor watch would signal the divers by rope to come up when an hour had elapsed (the USN paid five dollars an hour). When Jim and Verlan were discharged, he was put in charge of the diving locker by Lt. Cmdr. Dillon. Dives were later made in Guantanamo Bay. The USS *Missouri* went to Turkey and back, but there were no more dives made by this crew.

Sky and his wife, Madeline Fredrikson reside

in Worcester, MA. Together they have two sons, five daughters, 20 grandchildren, and 12 great-grandchildren.

CLARENCE C. FREEMAN JR., born Feb. 24, 1932 in Thomasville, NC, originally joined the Naval Reserves in 1948 and was called into the USN in 1952. After six weeks of training at Bainbridge, MD was assigned to the *Mighty Mo* and was in the supply division and achieved the rank of seaman first. After three weeks out of boot camp, they were on their way to Korea.

Memorable experience: they came to Japan for a rest and he had the opportunity of singing with the ship choral group in hospitals to those who had been wounded. He was on duty when the ships helicopter was shot down and all aboard were lost.

He was discharged in 1954 and awarded the China Service Medal and Korean Service Medal.

He married Carolyn Callicutt Freeman in 1956 and they have one son, a doctor, who has three children.

After service he graduated from High Point University, then taught school and coached football, basketball, and baseball for five years. He retired from Security Capital Bank in 1993.

HAROLD WOODROW FRENSLEY, born June 17, 1915 in Duncan, OK, enlisted in the USN Jan. 6, 1942, was a radioman and served in the Pacific and Aleutians 16 months. He returned to the States January 1944 and was assigned to the USS *Missouri* on Feb. 15, 1944.

He is a plankowner and served in CR Div. on the *Missouri*. He has four Battle Stars on his Pacific Ribbon and was present at the surrender aboard her in Tokyo Bay, Sept. 2, 1945. Upon his return to the states, he received his honorable discharge, Oct. 23, 1945.

He was employed by Atlantic Richfield Corporation for 28 years and lived to enjoy eight years of retirement. He died one week after attending the recommissioning in Long Beach.

He left Frances, his wife of 41 years, two daughters, Janeen and Alice, four grandsons and one granddaughter.

His family is very proud of him.

EVERETT NEWELL FROTHINGHAM, born Nov. 13, 1921 in Newburyport, MA, graduated Amesbury High School in 1936 and Tilton Junior College in 1941.

His father owned an airport on Plum Island, MA where he taught his son how to fly. He was accepted as an aviation recruit at Squantum Naval AB and received his wings and was appointed lieutenant (jg) July 1, 1943 at Jacksonville Naval Airbase.

He served aboard the USS *Missouri* from its commissioning June 11, 1944 until he was lost at sea due to a crash of his Seahawk SC1 aircraft on Feb. 10, 1945. He was 23.

He was survived by his wife, Regina Anne, a daughter, Sandra Regina, born Feb. 22, 1944 and a son, Lawrence Everett, born May 1, 1945, whom he sadly never saw.

His widow Regina Anne passed away Dec. 4, 1975 and is survived by their children and grandchildren.

In the town square of Salisbury, MA, a memorial plaque is dedicated to Everett Newell Frothingham. The plaque reads "Pilot - killed in the line of duty in the Western Pacific on 2/10/45".

Both Sandra and Lawrence are honorary members of the USS Missouri Association. There will always be an emptiness in the hearts of his children!

ALEX GABRYSZAK, born Feb. 6, 1927 in Buffalo, NY, inducted into the USN, assigned to USS *Missouri* June 1945, as powder man and primer man, turret #1, 1st Div.

His memorable experiences: sailing into Tokyo Bay and being a witness to the signing of the surrender document aboard the USS *Missouri;* traveling to New York City in October of 1945 to celebrate Navy Day and having President Harry S. Truman as an honored guest.

While at sea he participated in the last bombardment of steel mills in Murorah on Hokkaido.

He was awarded the Victory Medal, Asiatic-Pacific Medal, American Theatre Medal and discharged August 1946 as S2C.

He married Sally and they have three daughters and is retired from Calspan Corp. Gabryszak enjoys traveling and playing golf with his grandchildren.

VINCENT GALGANO, born July 5, 1918 in Brooklyn, NY, was inducted into the USN Jan. 6, 1942, and assigned to the USS *Missouri* February 1944-October 1945 as radio man.

While at sea he got the message that Japan was surrendering Aug. 11, 1944.

He was awarded the Good Conduct Medal, Pacific Theatre Award and discharged Oct. 23, 1945 as RM2C.

Galgano has been married to Joan for 50 years and they have one son and daughter. He is now retired from the post office after 38 years.

PASQUALE (PAT) GALLO, born Jan. 22, 1916 in Bridgeport, CT, joined the USN August 1943, assigned to USS *Missouri* March 1944, as water tender. His assignments and stations include: boot training at Sampson, NY; Norfolk, VA; Newport, RI; Brooklyn, NY; commissioning of *Missouri* June 1944.

He was stationed in boiler room on ship. One day a Japanese pilot, in a suicide attack, hit the ship. All at once smoke was coming through the vents and the doors closed. They always hoped that when this happened that the damage wouldn't be bad and they wouldn't have to flood the compartments.

While at sea he participated in the battle of Iwo

Jima and battle of Okinawa. He was discharged as WT3C, Nov. 16, 1945 and received a star for each battle.

He married Marie Berger and they have three children, son, Bob; two daughters, Carole Godfrey and Bonnie Johnson and six grandchildren: Kevin, Lisa, Gina, Kelly, Stacey and Michael and one great-grandchild due in August.

ROGER B. GAUTHIER, born in Rochester, NH on Sept. 27, 1920, enlisted in the USN on July 1942. He served aboard USS *Texas,* then the USS *Missouri.*

He has vivid memories of all the places the ship docked; a little opener made from pieces of the Japanese (suicide) plane; present at the signing and always thought of it as a historical experience.

His rating was S2c S1c Cox. He was discharged on Nov. 16, 1945. Gauthier passed away Oct. 23, 1995.

ROBERT A. GENEST, born Aug. 12, 1928 in Waterbury, CT, joined the USN Nov. 5, 1945 and was assigned to USS *Missouri* Aug. 13, 1946, A Div.

While at sea he participated in maneuvers off Cuba and Trinidad also Newfoundland and the Arctic Circle.

His memorable experiences: getting hit by 5" star shell while on maneuvers by USS *Little Rock;* one was killed.

He was discharged Oct. 21, 1947 as F1 and received the WWII Medal.

Genest has three children and three grandchildren. His wife Loraine died last September from brain cancer. He is now retired from the auto mobile business. He lives in Florida winters and Maine in summer.

CLARENCE (SKIP) GEORGE, born March 31, 1931 in Lackwanna, NY, was inducted into the USN July 16, 1948 and assigned to the USS *Missouri* Oct. 7, 1949 until discharge May 15, 1952 as machinist mate, #2 fireroom lower level.

His memorable experiences include January 1950, USS *Missouri* aground on Thimble Shoals in Chesapeake Bay for 15 days; 20 hour work days off-loading ammunition, stores, and fuel oil; August 1950, hurricane off Cape Hatteras enroute to Korea; September 1950, joined Task Force 95 in Korea in a typhoon; Christmas Eve 1950, protective bombardment for evacuation of UN troops at Hungnam.

George was awarded the United Nations Service Medal, Korean Service Medal w/3 Battle Stars,

Republic of Korea Presidential Unit Citation, Navy Occupation Service Medal, National Defense Service Medal and the Good Conduct Medal.

He married Mildred Ham and they have two daughters and three sons. He retired from Invest Financial Corporation and was manager of compliance.

EDWARD W. GODFREY, born Oct. 9, 1929 in New York City, enlisted in the USN Nov. 21, 1946. Served aboard the USS *Taconic* (AGC-17), USS *Providence* (CL-82), the USS *Missouri,* S-2 Div., 1949-51, US Naval Facility, London, England 1951-53. He received a BBA in accounting in 1960.

His most vivid memory is of the *Missouri* and other ships of the 7th Fleet covering the evacuation of the 1st Marine Div. from Hungnam, Korea in December 1950. The Marines had to fight their way to the sea after being trapped at the Chosin Reservoir near the Manchurian border.

Godfrey married Gloria Cardenas, has two sons, and a daughter. Retired from Federal Service in 1985 after 35 years total service. He has lived in Bowie, MD for over 32 years and is a volunteer in church and the community.

ANGELO (GUFFY) GOFFREDO JR., born Dec. 17, 1927 in Schenectady, NY, joined the USN July 1946 and was assigned to the USS *Missouri* October 1946, in electronics, radar-radio.

While at sea he was aboard when star shell hit 03 level, starboard side; 1946, President Truman presented silver star.

His memorable experiences include being aboard ship when it went aground in 1950; (President 1992-1996 USS Missouri Association) Office; USS Association 13 years; midshipman cruises to Europe.

He was discharged July 1950 as ET3 and is now retired as electrical contractor, active in USS Missouri Association. He married Rose and they have a son, Michael and a daughter, Georgianna.

LAWRENCE E. GOODMAN, born March 12, 1920 in Manhattan, entered USN May 18, 1943, worked on development of radio proximity fuse and its MK 64 blind-firing antiaircraft gun director. Joined USS *Missouri* July 1944 at Navy Pier, Bayonne, NJ following maiden voyage. Rank: ensign USNR (Ltjg.) Assignment: supervision of installation, maintenance and crew instruction for four MK 64 directors controlling five-inch/38 gun mounts. Detached March 1945.

He received the American Theatre, Asiatic-Pa-

cific w/2 stars and Victory Ribbons. He is proud to have been a small part of the initial combat actions of the finest ship and crew in the USN.

Goodman married Katherine C. Lewis in 1951 and they have lived happily ever after. They have three lovely daughters and three granddaughters. He taught at Columbia University, University of Illinois, University of Minnesota, Cambridge University (England) and Texas A&M University, published two books and received the Newmark Gold Medal of the American Society of Civil Engineers. Nothing that followed displaces the memory of the fine performance of the 5"/38 guns and director crews at the demonstration in Ulithi lagoon held for the chief gunnery officer of the Pacific Fleet. Afterwards, most of the antiaircraft shells fired at Okinawa were armed with radio proximity fuses.

JAMES B. GOSNELL SR., born Feb. 6, 1922 in Pelzer, SC, attended Pelzer schools, graduated in 1939, enlisted in USN in August 1940. Attended boot camp in Norfolk, VA. First ship USS *Mississippi* (BB-41) 1940-1942. In Pacific and North Atlantic. He then was plankowner in USS *Columbia* (CL-56) 1942-1946. Participated in operations from Guadalcanal through Philippines and anti-shipping sweeps in the East China Sea in 1945. The ships action included shore bombardments, enemy air attacks (three kamikazes hit the *Columbia*) and in the last two major surface engagements of the war.

He was a tug master 1947-1949 at Charleston Naval Shipyard.

1949-1954 on board the USS *Missouri* (BB-63). He was the leading boatswains mate of the 6th Div. In 1952 he was advanced to chief boatswain's mate. The *Missouri* had two eight month deployments to Korea 1950-1951 and 1953-1954.

After *Missouri* he was the chief master at arms of the NAS Guantanamo Bay, Cuba 1954-1957. From 1957-1959 he was the Navy recruiter in Tallahassee, FL.

In 1959 he left for the Persian Gulf to work with hydrographic surveying in the USS *Tanner* (AGS-15) as boat officer of a 40 foot sound boat. He gained more valuable experience operating independently 10 days to two weeks at a time.

In 1960 he was ordered to Mine Squadron Ten in Charleston, SC. There he worked with mine countermeasures and was commander of Mine Div. 102 and operations officer for the squadron.

In 1965 he was assigned to the USS *Ozark* (MCS-2) as the mine warfare officer in charge of 20 36' mine sweeping boats and the mine sweeping gear.

He retired as a lieutenant with 28 years service. He worked with Metropolitan Life Insurance Company for 16 1/2 years and retired again. He and his wife, Betty Freeland Arnold Gosnell have six children: Frank (Z) Barbaras, Skip, Elizabeth and Jennifer, all this and only four grandchildren. He resides in Charleston, SC.

ELWOOD H. GRAFF, born June 17, 1928 in Brooklyn, NY, was at the launching of the *Mighty Mo* on Jan. 29, 1944 as a teenager. His uncle was a master carpenter at the Brooklyn Navy Yard and he invited him to see the launching.

He enlisted in the USN on Dec. 27, 1945 and took boot training in Norfolk, VA. He boarded the USS *Missouri* on May 9, 1946 and served in the A Div., forward diesel, making all the fresh water that was used aboard the *Mo* from the laundry to the galley to the

showers and to the ship boilers. Emergency diesel generators were also maintained in forward diesel.

On Sept. 7, 1947 the USS *Missouri* picked up President Harry S. Truman along with his wife, Bess and their daughter Margaret in Rio de Janeiro for their return trip to the USA. President Truman and his family had flown to Rio for the signing of the Rio Treaty. It was a memorable trip because the crew held the shellback ceremonies on the way back so that the President and his family could participate. It was understood that this was the first time that a woman had ever crossed the equator on a US Man-of-War.

Graff was discharged Nov. 25, 1947 at the Flushing Avenue Barracks in Brooklyn with the rate of MM3/c and received the Victory Medal.

He married Jeanette Gorback in 1950 and they have four children and 13 grandchildren. He retired from the postal service in 1989. "We Remember"

LEONARD W. GREATHEAD, born Sept. 27, 1920, enlisted in the USN on June 13, 1940. Served on USS *Boise* as gunnery instructor at Fleet Service Schools, Norfolk, VA and USS *Augusta*. Pre-commission detailed to USS *Missouri* May 1944 and plankowner. As BM2/c was police petty officer, 7th Div., mount captain, Mt. #7, 5"-38 antiaircraft guns.

He saw a kamikaze hit the USS *Princeton* aircraft carrier. Their ship engaged in shore bombardment of Okinawa enabling them to take Shuri Castle, the island fortification.

A bonus for six years' service was a "pleasure" tour to Turkey, Athens, and an audience with Pope Pius in Rome.

He was awarded the Asiatic-Pacific Area w/5 Stars, Good Conduct Ribbon w/Star, American Theatre Ribbon Victory Medal and American Defense Ribbon w/Star.

Greathead was discharged July 10, 1946. He married Marie Marks in Washington, DC and their family includes a daughter, Colleen, her husband Gary Voet, grandson, Brian, daughter Sheila, her husband, Bruce Scheidt, grandson, David and granddaughter, Courtney. He retired as service technician from Bay Alarms, a burglar and fire alarm company.

FREDERICK W. GREAVES, born Sept. 28, 1930, entered the USN Oct. 2, 1947 and was assigned to the *Missouri* March 3, 1948 and discharged Sept. 10, 1951.

His memorable experiences include: first and above all, meeting Glen M. Stewart from Dubois, PA, who became his best friend for life; meeting President Harry S. Truman. He came into Greaves work station

one day. He doesn't remember what year 48, 49 or 50. They got to talking and the President told him to call him Harry. Being a young kid and not very worldly, Greaves said to him, "Harry you are a great guy, but I don't think I could vote for you." The President said, "Why is that? Greaves said, "Because my family are all Republicans and I'm going to vote that way." The President said to Greaves, something he would never forget. "Son this is what makes our country so great, you can be wrong and I'll still love you." He will be ever so grateful to the *Missouri* for giving him a chance to meet such a wonderful human being. Also while in Korea, Christmas time in 1950, he had an opportunity to meet Bob Hope, another wonderful and caring person.

While at sea during the early part of the war at Inchon Landing, he engaged in many shore bombardments including the Hungnam evacuation.

He has been married to the former Roberta Cortez for 43 years and they have three wonderful children and four grandchildren. He is now retired from a business that he and his son Fred started. His son is running the business himself and doing a great job. He is also taking good care of his mother and father.

GEORGE L. GREER, born May 30, 1925 in Columbus, OH, enlisted in the USN Aug. 4, 1942 and was sent to Great Lakes NTS. After boot camp he was assigned to the USS *New York,* then participated in the invasion of North Africa at Casa Blanca, where during operations, they were fired upon by the French battleship, *Jean Bart.*

He did convoy duty in the North Atlantic after North Africa operations. He was assigned to the *Missouri* crew in April 1944 at Newport, RI NTS. Greer is a plankowner. He went aboard the USS *Missouri* June 1944 and discharged June 6, 1946 as S1C

Greer's memorable experience includes being present at the surrender ceremony. He was awarded the ETO w/star, Asiatic Pacific w/3 stars.

He is married to Jean D. Greer and they have two sons, one daughter and seven grandchildren. He retired from Rockwell Institute and is playing golf and enjoying life.

MICHAEL GREGORY, born Aug. 29, 1927 in Boston, MA, enlisted in the USN April 1945 (two months before graduation from high school).

He served in Sampson, NY for boot camp; Bainbridge, MD, for Hospital Corps School; Portsmouth, NH Naval Hospital and USS *Missouri* until discharged in August 1946 as a PhM3c.

Gregory graduated from Northeastern University and Brown University (thanks to the GI Bill). Employed simultaneously as a registered pharmacist and chemistry teacher for 25 years. Currently employed part-time as a plastics chemist. He married Gloria Catrambone and they are the parents of two sons, two daughters and 14 grandchildren.

He has fond memories of the *Missouri's* cruise to Greece and Turkey, maneuvers in the Caribbean; visitors day in New York; Hatuey beer, a chocolate

drink in the ship store; a baseball game with the *Wisconsin*; the movie *Gilda* with Rita Hayworth and the projector breaking down just before the interesting part, when she was singing, "*Put the blame on Mame . . .*"

JAMES C. HAMPTON, born Sept. 12, 1924 in Quitman, GA, enlisted in the USN, March 3, 1944. After training at Bainbridge, MD and Providence, RI he was assigned to the USS *Missouri* June 1944. After loading supplies, they headed for Pearl Harbor for a half day liberty, then on to join the Pacific Fleet. Participated at Iwo Jima from February to March 1945, at Okinawa, May 1945, and the bombardment against the Japanese home islands August 1945. Also served on the USS *Fitch* (DMS), USS *Albany* and Guam. Taught fire fighting on Treasure Island for four years.

Hampton was discharged April 1, 1965 as DC1 with authorization to wear the Pacific Ribbon w/3 Stars, American Campaign Medal, Naval Occupation Service, five Good Conduct Medals, National Defense Service, WWII Victory Medal, Philippine Liberation and the Philippine Independence.

He proudly attended the surrender signing ceremony of Japan to the USA on the deck of the USS *Missouri* on Sept. 2, 1945.

He is married to Louise and they have four sons, two daughters, and five grandchildren. They are now living in Phoenix, AZ.

JAMES M. HAPENNY, born Nov. 28, 1929 in Newton, MA, joined the USN Dec. 26, 1946 and was assigned to the USS *Missouri* March 7, 1947. He served as FC in the main battery.

While at sea he participated in Korea and 52 consecutive days of condition 1, 2 and three.

His memorable experience includes meeting General MacArthur and President Harry Truman.

Hapenny was discharged as FC3 Dec. 12, 1951 and awarded the Good Conduct, Korea and United Nations Ribbon.

He is retired from the Raytheon Company after 40 years and is married to Geri. They have sons, Jim and Mark and daughters: Darlene, Tammy and Laura.

GEORGE A. HARRIMAN, born Sept. 1, 1929 in Kennebunk, ME, enlisted in the USN Dec. 18, 1946 and was assigned to the USS *Missouri* Spring of 1947. He served in the 5th Deck Div., gunnery, lower handling room 5"38.

His memorable experiences include the Sept. 11, 1947, equator crossing; carrying President Truman and family northward bound.

While at sea he participated in the Blood Island bombardment, no combat, but 16" rifle salvos were awesome. He was awarded the Victory Medal and the Occupation Service Medal. He was discharged Sept. 25, 1951 from a second enlistment.

Harriman is married to Shirley Libby, his second marriage, and has six children, 13 grandchildren and one great-grandchild. He retired after 37 years in Civil Service career.

ALVIN RICHARD HARRIS, born June 14, 1924 Newport News, VA, joined the USN December 1942 at the University of Virginia, Charlottesville, VA. V12 trained at University of North Carolina June 1943-June 1944. Sent to Asbury Park, NJ for further training. Sent to Cornell, Ithaca, NY for Midshipman School and commissioned January 1945 as an ensign. Sent to Gulfport, MS for further training as a recognition officer. First sea duty assignment was to the USS *Franklin* (CV-13) but on the way there the *Franklin* was damaged and he was reassigned to the USS *Missouri* (BB-63) flagship of the 3rd Fleet (Adm. Halsey). He was a gunnery officer in the 3rd Div. and received two Battle Stars for the Kyushu Campaign and Mainland of Japan Campaign and the Good Conduct Medal.

September 1, 1945 (BB-63) behind the minesweepers sailed into Sagami Wan and then into Tokyo Bay. Thousands of Japanese civilians dressed in black and white lined the banks to see the mighty US fleet arrive that night. It was very interesting. September 2, 1945 the Japanese surrender ceremony and it was the 3rd Div. that brought the Japanese aboard to meet with Gen. MacArthur. They sailed to Guam under Adm. Chester Nimitz as his flag, then on to Pearl Harbor and Panama City. Through the Panama Canal and on to New York for Navy Day October 1945. President Harry Truman and his family came aboard for lunch with the officers and to address the crew. They were then ordered to Guantanamo, Cuba for maneuvers. February 1946, under the Truman Doctrine, they sailed for the Mediterranean and returned the body of the Turkish ambassador, to Istanbul. He had died in the US during WWII.

They made many other ports in the Mediterranean, all were memorable meetings with dignitaries. They paraded, they were decorated, wined and dined wherever they went. They met with Pope Paul in his chambers then went to Gibraltar and back to the US.

Harris was separated June 1946, graduated University of Virginia in 1948. Founded his own business February 1949-October 1990. He retired at age 65. He has been married for 40 years to Joan Blumenthal. They have four children, three granddaughters and live in Avon, CT.

LES HARRIS, born July 16, 1932 in Newburyport, MA, was inducted into the USN March 20, 1952 and assigned to the USS *Missouri* February 1953. He served as musician, second loader on twin 40s.

His memorable experiences include initiating shellback and crossing the equator, July 1953.

He was awarded the National Defense Medal and discharged May 4, 1954, MUS.

Harris is the father of four and has five grandchildren. He is a retired professor from Berklee College of Music, Boston, MA. He has been drummer for the late Bobby Hacket, the late Zoot Sims and Buddy DeFranco.

ROY JAMES HARTMAN, born June 11, 1932 in Pigeon, MI, enlisted in the USN, April 21, 1952. Attended boot training (11 weeks) at Great Lakes, IL. August 5, 1942 transferred to USS *Missouri* at Norfolk, VA for duty. Assigned A Div. forward emergency diesel generator room and fresh water evaporators. In port duty, boat engineman. September 1952 USS *Missouri* departed for Korean combat zone. Returned to the States April 1953.

He was married May 20, 1953 at Pigeon, MI. June/July midshipman cruise to Rio de Janeiro, Brazil. Crossed equator June 19, 1953. May 17, 1954 transferred to Great Lakes Engineman Class "A" School (14 weeks). October 5, 1954 returned to duty on *Missouri* at Bremerton, WA. Ship being decommissioned. November 16, 1954 advanced to Eng. 3. December 20, 1945 transferred to USS *Oriskany* (CVA-34) at Alameda, CA. Assigned to A Div. after emergency diesel generator. In port duty, duty boat repairmen. Departed for Far East. Returned July 1955. November 11, 1955 advanced to Eng. 2. February 1956 left for Far East. March 9, 1956 transferred receiving station, Honolulu, HI. March 12, 1956 transferred via air transport to Treasure Island, honorable separation March 16, 1956.

His memorable experience: While visiting the Navy hospital in Yokosuka, Japan and talking about the *Missouri*, a young Marine came up. He asked "Did you say you were off the *Missouri*?" Harris said, "Yes." He held out his hand and they shook hands. He told Harris that he was grateful to the *Mo*. His patrol was pinned down, too far for their artillery to help and weather was too bad to have planes come to them. Then over their radio came "this is the USS *Missouri*, can we be of some assistance to you?" They gave their position and the 16-inch shells started whistling over their heads. What a wonderful sound! It gave them time to withdraw. Still shaking Harris's hand he said thank you. This made it all worthwhile.

His campaign ribbons include the Korean Service, United Nations, National Defense, Good Conduct and China Service.

He spent a year in Los Angeles, CA after serving, then returned to Pigeon, MI. He has been married to Margaret for 36 years and they have two children and three grandchildren. They are insurance agents and owners of an independent insurance agency in Pigeon, MI.

ROBERT T. HEFTY, born Feb. 27, 1932 in Fargo, ND, moved with his family to Valley City, ND in 1941. Grew up, attended schools and graduated from College High School there in 1950. Attended V.C. State College for a time and enlisted in USN April 28, 1952. He went through boot camp at Great Lakes NTC. Assigned to USS *Missouri* in August 1952. Then assigned to the E Div. where he was a electrician's mate, working in distribution. Later attended 14 week "EM" "A" School in Great Lakes and eventually attained the rank of EM1c.

The *Missouri* departed Norfolk in September 1952 for Korea. While in Korean waters, the *Missouri*

engaged in shore bombardments, ground troop support evacuation and good will missions.

One experience he will always remember is the day in 1952 when the word was passed on the 1 MC had just participated in setting a naval record; the longest siege of any one port. Two years of constant bombardment of Wonsan Harbor; four days, riding out an Atlantic hurricane, out of Norfolk in 1954. *Missouri* took it real well. Hair raising time though. *Missouri* also supported friendly forces, North, by the Yalu River area.

After their last day of action in April 1953 they were entering the harbor of Sasebo, Japan to resupply for their trip back to stateside, their Skipper, Capt., W.R. Edsall suffered a heart attack and died on *Missouri's* bridge, "in command".

They received the National Defense Service Medal, Korean Service Medal w/2 Battle Stars, China Service Medal, United Nations Service Medal and Korean Presidential Unit Citation, for their action in Korea.

The *Missouri* returned to Norfolk, later, sailed to Bremerton, WA and decommissioned in February 1955.

Hefty was transferred to the minesweeper, USS *Endurance* (MSO-435), out of Long Beach. He received the Good Conduct Medal, while aboard the *Endurance*.

He was separated from service April 1956 and returned to Valley City, married Sonja Overbo and took a job as electrician for V.C. Public Works and worked there for 37 years. He and Sonja have two sons. Rocky married (Laval) and Alan is single. They have no grandkids yet.

He was totally disabled by a stroke in 1993 and turned 65 in February. He and Sonja are now on the pension list. They recently bought their retirement home, so that keeps them busy and out of mischief.

Hefty is a member of the USS Missouri Association, American Legion, a life member of both the American Battleship Association and the VFW.

KENNETH J. HEIDER, born June 10, 1922 in Johnstown, PA, entered the USN May 11, 1942 serving on the USS *Cowie* (632) in the North Atlantic. Attended diesel and watertender school and was assigned to the USS *Missouri* at Brooklyn, NY navy yard in 1944. Stayed at the Flushing Ave. barracks until the living quarters were ready on the *Missouri*. He traced all the fuel, oil, water and steam lines through all four engines and fire room. They went on a "shakedown" run to the Port of Spain, Trinidad; participated in other exercises before putting the USS *Missouri* in commission.

While going through the Panama Canal to the Pacific they used a 12-inch ruler to measure the distance between the sides of the *Missouri* and the canal to keep the ship from rubbing.

After the Japanese surrender he witnessed the signing of the Peace Treaty in Tokyo Bay.

He married Margaret (Peg) Hahn June 29, 1946 and they have two daughters, two sons, three grandchildren and one great-grandchild.

Heider is currently active in the VFW and has held positions as post commander (Post 6233), County Council Commander and District 28 Commander; as well as all chairs leading to the commander positions.

BRODUS W. HINSON, born Feb. 24, 1928 in Anderson County, SC, joined the USN Feb. 16, 1946, assigned to the USS *Missouri* March 1946, gunnery 6th Div., achieved the rank of S1c.

His memorable experiences include transporting President Harry Truman and family from Brazil to the States; the menu from Thanksgiving Day 1947 on the North Atlantic Ocean.

He was discharged Dec. 10, 1947. He is married to Barbara "Bobbie" McMullan and they have three sons (youngest deceased at 17) two daughters and seven grandchildren. Hinson retired as a brick mason and construction superintendent and is enjoying life on Lake Hartwell, SC.

HARVEY C. HINSON, born March 27, 1926 in Salisbury, NC, enlisted in USN Aug. 3, 1943, eighth grade education. He later went back to finish. Stationed at Norfolk, VA, USS *Chilton,* Newport, RI and Brooklyn, NY where the USS *Missouri* was constructed and put in commission. He was on the *Missouri* until discharged March 23, 1946 in Lido Beach, Long Island, NY.

There is a lot to remember. A lot of things happened. One was when they were hit by a Japanese suicide plane. After the fire was put out, the machine gun from the wing of the plane was sticking through the barrel of one of the 40mm (like sticking a tooth pick through your fingers).

He was discharged a SF2c and received the American Theatre Medal, Asiatic-Pacific Medal w/3 Stars, and Victory Medal.

Hinson is married to Irene McKinney and they have one daughter, Tamara J. Hinson Barker, and one grandson, Andrew Barker. He retired from the VA Medical Center, Salisbury, NC after 35 years in the Engineering Department Dec. 30, 1988.

HENRY P. HOBERG JR., born Sept. 19, 1930 in Catskill, NY was inducted into the service May 25, 1950. He had his training in Great Lakes. From June until mid August they were told of a 600 draft to serve on the USS *Missouri*. They traveled to Norfolk, VA for two or three days to get supplies to be put aboard ship and left Norfolk to embark on a trip to Korea, by the way of the Panama Canal and stopover in Hawaii of four days.

Their first big action was an invasion of Inchon. It was getting lonely not having mail from home, and one day while group of them were on watch, Hoberg met another sailor that had a magazine of the *Youth for Christ.* He got the address and had his name listed in it. Before Christmas of 1950 he got mail from a girl in Lancaster. They wrote back and forth as pen pals and then met in November 1951.

He had to go to Korea for another six months and was getting serious about his pen pal. He asked

for shore duty on the East Coast. They were engaged in 1953 and he was discharged in March 1954. He married Ruth Thomas May 2, 1954 and they have lived in Lancaster, PA ever since. They have five children: Steve, Susan, Debbie, Dave and Robert. Hoberg worked for the Buick Company from 1976 until his retirement in 1992. He met two truck drivers at the Buick Co. and they told him about the USS Missouri Association. He has been with the association since attending his first meeting in Nashville, TN in 1986. He has met sailors from WWII and Korea.

LARRY HOGAN, born Jan. 31, 1921 in Chicago, IL, was assigned to the *Missouri* "H" Div. in Newport, RI where they were forming the crew in 1943. He stayed on the *Mo* until he was discharged in 1946.

He had many great experiences but the greatest and most exciting occurred when it was announced that they were going to New York for Navy Day. That meant going home; you can't beat that for excitement.

Hogan is married to Doris and they have four children.

LOWELL G. HOLMES, born Feb. 10, 1927, was inducted into the service April 21, 1946. He went aboard the *Mighty Mo* July 1946 and served until March of 1948.

One things that stands out in his mind about the time he served on the *Big Mo* occurred when they thought they had a fire under the deck of one of their 16 inch powder magazine, the paint was curling up on the deck under the cans of powder and they knew they had only a few short moments to do something. The XO was there to direct the operation of the fire fighting, and he told another sailor to go get the key to the hatch going down into the bilge under the magazine. Another sailor named Buchannen from North Carolina told Holmes to get hold of the other side of the hatch handle and he would unlock the lock. Holmes reached down with one hand against the bulkhead, the same as Buchannen did and they jerked up on the hatch and broke the lock as clean as if it was cut. That is how frightened they were. When they got under there they found a ruptured steam line and were thankful. No one could ever believe how they broke that lock, but no one knew how frightened they were either.

He was honorably discharged March 12, 1948 as S2c and received the Service Medal and Victory Medal.

Holmes is married and has four children: Neva Ransom, Don Holmes, Patricia Tweedy and David Holmes and nine granddaughters. He retired from Monsanto Chemical Company and is now in the heating and air conditioning business.

AMORY M. HOUGHTON III, born Dec. 31, 1929 in Bangor, ME. Upon completing high school, he enlisted in the USN Aug. 23, 1948. Attended boot camp and fire control school at Great Lakes, IL. Boarded the USS *Missouri* April 1949 in Norfolk.

Assigned FA Div. Released to active reserves August 1949 after a short tour on USS *Mississippi*. Recalled August 1950 for Korean War duty. Reassigned to *Missouri* and FA Div.

Support of troop evacuation at Hungnam Christmas 1950 was his most vivid memory. Quite an accomplishment. Inchon, Wonson, Chong Jin, indelible recollections. He received the National Defense Medal, Japanese Occupation Medal, United Nations Service Medal, Korean Service Medal w/3 stars and the Korean Presidential Unit Citation.

Houghton left active service as FCS3 May 1952 and was discharged August 1953. He graduated from Cornell University. His first wife, Martha Beith died and he remarried Joan Paradis. Together they have six children and five grandchildren. He retired February 1991 after 30 years with Portland (ME) Press Herald as business and facilities manager.

JAMES E. HURST, born June 6, 1944 in McKeesport, PA, joined the USN April 20, 1973 and was assigned to the USS *Missouri* September 1989. He served in OA Div. as LPO weather forecaster and computer center; damage control central coordinator; missile key holder; OOD and JOOD duties; senior first class PO president; first class mess (1990-Decom).

He visited Matzatlan, Hawaii, Seattle, Vancouver, San Francisco, San Diego, Subic, Phattaya Beach, Bahrain, Perth, Hobart Desert Shield, Desert Storm, 50th anniversary, Pearl Harbor.

His awards include: Persian Gulf; two combat actions, SW Asia, Kuwait Liberation, Expeditionary, Sea Service, Saudi Arabian Medal, Humanitarian, Navy Battle E, Good Conduct (5), Surface Warfare and Air Warfare Specialist.

Hurst was medically retired April 30, 1993. He is married to Barbara and they have one daughter, Susan. They reside and are actively involved in the community of Springfield, MO. His organizations include: Shrine, Masonic, Elks, Fleet Reserve, American Legion, VFW, DAV, AA, Cancer Support Group, and American Red Cross, Disaster and Military Family Services plus CPR and 1st Aid instructor. Cable TV Commission. Colon cancer survivor.

ROBERT J. HUSSEY, born Jan. 2, 1934 in New Haven, CT, enlisted in the USN Nov. 7, 1951. He received a high school education and reported aboard the USS *Missouri* with 7th Div., March 15, 1952 at NOB Norfolk, VA. After midshipmen cruise to Norway, England, Cuba and the Far East with the 7th Fleet, winter 1952-1953 (Korea) and second midshipmen cruise to Rio de Janeiro. The summer of 1953 he transferred to USS *Wisconsin,* 7th Div. and returned to Far East. Discharged Oct. 26, 1954 at T.I. San Francisco, CA.

Hussey went to work for Sikorsky Aircraft Div. of United Tech and retired in 1989 after 35 years.

He married Jo Ann Hamilton Feb. 8, 1958 at the Sacred Heart Church in Newhaven, CT. They have two daughters, one son and two granddaughters. Thanks to their association they are able to meet and greet ship-mates after 40 plus years.

JACK HYNES, born Dec. 28, 1926 in Holyoke, MA, enlisted in the USN December 1943. He attended boot camp and Signal School in Sampson, NY. Served in USS *Spector* (AM-306); USS *Terror* (CM-5); USS *Missouri* (BB-63). Received the Purple Heart for wounds received on USS *Terror,* May 1, 1945.

His memorable experiences include minesweeping off Iwo Jima, Feb. 16, 1945; observing first flag raising on Mount Suribachi (Hot Rocks) February 23 in a.m; USS *Terror* answered general quarters 93 times in April; Ship hit by kamikaze May 1, 1945; USS *Missouri* Mediterranean cruise 1946.

He was awarded the Purple Heart, American Campaign, Asiatic Campaign w/2 Battle Stars, Victory Medal, Occupation Medal, Philippine Liberation, Independence and Presidential Unit Citation.

He is retired after 30 years as motion picture exhibitor and is married to Janet MacKenzie. They have a son, Johnathan; daughter, Cynthia Hynes Claydon; a grandson, Sgt. James Claydon, US Army and two great-grandchildren.

CLIFFORD HYSTAD, born April 17, 1929, joined the USN Sept. 14, 1948 and assigned to the USS *Missouri* Nov. 20, 1948, 5th Div.

He served in Korea from August 1950 to March 1951. His awards include the Good Conduct Medal, Navy Occupation Service Medal w/Asian Clasp, China Service Medal, National Defense Service Medal, Korean Service Medal and UN Service Medal w/star.

While serving aboard the *Mo* he had the unique experience of having liberty during the invasion of Inchon.

He was running a motor whale boat delivering guard mail to the USS *St. Paul* when someone called down from a merchant transport. He looked up and saw his brother. He asked the boat officer if he could stop for a few minutes. Hystad found out he was going in about three hours. After returning to the *Mo* he asked the XO for liberty and told him the situation. The XO called a boat for him and Hystad spent two hours with his brother before he went in.

He is a retired farmer, has been married for 45 years and has three children and 10 grandchildren.

WILLIAM D. ISAACSON, born May 5, 1925 in Brooklyn, NY, was inducted into service Oct. 10, 1943, NRS, New York, NY. Attended boot camp at NTS, Sampson NY. Entered Signalman School, Sampson, NY and graduated as honor class signalman. Sent to Submarine Chaser Training Center in Miami, FL. Assigned to recruiting station, Jacksonville, FL; naval operations training command boat facility, Naval Auxiliary Air Station, Mayport, FL; ComF Carrier, Task Force Pacific, aboard USS *New Mexico;* USS *Enter-*

prise; USS *Missouri;* USS *Wisconsin;* USS *Guadalcanal;* ComBatCruLant (Commander Battle-ships and Cruisers of Atlantic Fleet).

He was discharged April 16, 1946 from PSC Lido Beach, Long Island, NY. Served on Flag Staff Commands of Adm. Sherman and Adm. Fechteler.

He was awarded the American Theatre Medal, Asiatic-Pacific Medal and the Victory Medal.

His most memorable experiences: bringing back the first load of 500 Marines on board the USS *Guadalcanal* after Japan surrendered; crossing through the Panama Canal.

Isaacson is married to Sue and they have a son, Brian, daughter, Beth and grandson, Justin. He is now retired.

ROBERT N. JACKSON, born in Woodford, WI, May 19, 1923, joined the USN Jan. 2, 1943. He took boot training at Farragut, ID and transferred to San Diego NTS for schooling, then to Mare Island Navy Yard for optical school. He was assigned to the USS *Missouri* October 1944 and was in the FM Fire Control Div. as SAO2c in charge of all optical equipment aboard ship.

His memorable experiences include a kamikaze hit April 11, 1945; damage in typhoon; signing of surrender in Tokyo Bay and returning to New York on Oct. 24, 1945.

He was discharged in November 1945. In 1946 he joined NP Benson Optical Company as a lens grinding technician. Then was employed at Sears Roebuck and Company. He went to plumbing, heating and air conditioning school for Sears and was there 17 years and later joined a wholesale plumbing and heating company.

Jackson married Jessie in 1948 and they have no children. He retired in 1985.

RUSSELL F. JAMES, born March 17, 1944 in Fall River, MA. After receiving a bachelors of fine arts from Southeastern University of Massachusetts in 1967, he entered USNR Dec. 20, 1967 and was stationed at the following places: Norfolk, VA; USS *Amphion* (AR-13) in 1968 to 1969; Portland, ME; USS *Exploit* (MSO-44) 1978 to 1982; USS *Mt. Baker* Charleston, SC 1982 to 1984; Long Beach, CA USS *Missouri* (BB-63) 1985 to 1987 plankowner; Long Beach, CA mobile inshore underwater warfare unit 105 1987 to 1991. Veteran of Desert Shield/Desert Storm 1990 to 1991. James was honorably discharged Aug. 18, 1922 as GMG1.

James states his most memorable experiences include: being part of GS Div. for first firing of 16 inches guns in almost 30 years on the *Missouri;* around the world cruise 1986; naval review 1986 Sidney, Australia; receiving his Golden Shellback. Russell James completed 17 years Reserves before being medically discharged after serving in Desert Shield/Desert Storm. Six of the 17 years Reserves were served in active duty.

He was awarded the National Service Medal w/star, Kuwait Liberation Medal, Naval and Marine Corps Overseas Service Ribbon and Sea Service Ribbon.

He is married to Tina Marie James and they have two daughters. He is currently retired and enjoying the Big Sky Country of Montana.

THOMAS J. JANSEN, born Aug. 30, 1925 in Cincinnati, OH, entered the USN Nov. 11, 1943. His stations include: Sampson, NY; Wentworth Institute, Boston, MA; USS *New York; Missouri,* MM2c, A Div. (fwd diesel room; evaporators, air compressors, April 1944 until discharged April 12, 1946 (plankowner).

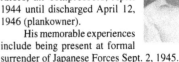

His memorable experiences include being present at formal surrender of Japanese Forces Sept. 2, 1945.

Jansen received the American Theatre, Asiatic-Pacific w/3 stars and Victory Medals.

He was awarded a BS and MSW degree. Retired February 1990 as state of Florida hearings officer. He is married to Elizabeth R. Jansen and they have four children and 11 grandchildren.

HENRY JONES, born Sept. 17, 1957, North Kingstown, RI. Naval Academy class of 1979. Stations: USS *Bradley* (FF-1041), NAVOCEANCOMCEN/JTWC Guam, OIC NAVOCEANCOMDET Kadena, NRL Washington, DC and CNO (N096). Served aboard *Missouri* as ships oceanographer/OA Div. officer from November 1989 to December 1991.

His memorable experience: all the warships at Naval Station Long Beach using their whistles to pay an emotional salute to the *Missouri* as she entered the harbor at sunset after completing her final voyage before decommissioning.

Surface Warfare and Navy Parachutist Designators. Awarded Meritorious Service Medal, Navy Commendation Medal, Navy Achievement Medal, Combat Action Ribbon, Navy Unit Commendation w/3 stars, Navy E Ribbon, National Defense Service, SWA Service w/2 stars, Saudi Arabia and Kuwait Liberation Medals, Sea Service Deployment Ribbon w/2 stars, Overseas Service Ribbon w/2 stars and the Navy Recruiter Gold Wreath.

He is currently a military instructor/oceanography lecturer at the Naval Postgraduate School, Monterey, CA.

A.E.W. (AL) KELLEY, a third generation Californian born Dec. 12, 1921 in Oakland. Accepted UC Berkeley fall semester 1939. March 18, 1942 accepted for Marine Corps Reserve as private first class. Graduated from California June 5, 1943, BS in forestry. July 24, 1943 active duty boot camp Parris Island, SC; OCS Quantico, VA. Commissioned second lieutenant, sent to Sea School, Portsmouth, VA. Assigned Marine Detachment USS *Missouri*. On the *Mo* shared a "stateroom" in the forward tower with Ens. "Ted" Harbert from Piedmont, CA which is surrounded by Oakland. Big ship, small world! Took part in commissioning ceremonies ergo a plankowner.

Prior to "shakedown" had dinner with a blind date Ens. Mary V. (Jini) Pierce USNR (WAVE) aboard the *Missouri*. After shakedown engaged to Ens. Jini Pierce.

General quarters: First, up in "Sky Control" had 20 mms manned by Marines. Second, 40mm quad atop turret three with an all USN crew. A good group of sailors. He liked that job. Third, down to CIC to coordinate the USN radar with the USN five inch guns. Finished the war at this post.

August 25, 1945 became "flag lieutenant" to Adm. Halsey by letter from Adm. Robert B. Carney his chief of staff. He was ordered to search the Japanese pilot (a four stripper captain) and his aides. On surrender day, Sept. 2, 1945, was acting CO of Marine Honor Guard. He is visible for a couple of seconds in *Victory at Sea*!

Back in US married Jini, the original blind date. The date was Oct. 29, 1945 with her father an Episcopal priest officiating at his church in Philadelphia, PA. Went on inactive duty March 6, 1946. Returned to Oakland. (promoted to captain in the reserves during the Korean War). They have three kids and three grandkids. He has a career in mortgage lending and established his own company. Now semi-retired and is still married to his Jini.

KENNETH K. KITTOE, born Aug. 13, 1928 in Aurora, IL, enlisted in USN Sept. 25, 1946 with a high school education. He trained at Great Lakes, IL and Armed Forces Staff College in Norfolk, VA. He was assigned to the USS *Missouri* 1948 to 1951. He achieved the third class boatswain, 5th Div.

There are so many memories. One was when they went aground in Hampton Rhodes, VA. It took about 17 days to get the ship freed. Another time was when they were on their way to Korea, they were hit by a hurricane and lost everything top-side. They stopped in Hawaii for repairs before they could go to Korea.

He married Marie Hart in 1953. Their family includes: Kenneth K. Kittoe Jr. and his wife Dawn, grandchildren, Kurin and Keith Kittoe; three daughters, Kimberly Johnson and grandson Jerad Johnson; Karla Sutton and her husband, Lester; Kelly Dolan and her husband, William and granddaughter, Sophia Dolan. He retired from AT&T, Montgomery, IL after 30 years in the maintenance department November 1986.

HARRY J. KLOSS, born Jan. 11, 1924 in Milwaukee, WI, served as a deckhand on USS *Henderson* and USS *West Virginia*. He left the *Virginia* one week before Pearl Harbor for Asiatic duty.

He was on the USS *Otis,* submarine tender then to the USS *Missouri* where he was an electrician on the No. 1 turret. He went to serve on the USS *Missouri* June 1944 until end of WWII and was aboard for peace signing. Aboard the USS *Otis* torpedoes were aimed at them, but missed. At the Cavite Naval Yard, Philippines they were bombed by Japanese, this a couple days after Pearl Harbor. Lost his hearing then, but stayed in service until the end of the war.

He was awarded the American Defense w/star, Asiatic-Pacific w/4 stars, American Area, Philippine Defense and the Good Conduct Ribbon.

Kloss retired from the Milwaukee Public Schools, worked in maintenance, playgrounds, fences, stadiums, etc. He married Agrapine and they

have a daughter, Lanice, son Lamar and three grandchildren.

LARRY L. KNIGHT, born July 9, 1966 in Coffeyville, KS, enlisted in USN May 1984. Stations include: USS *Yellowstone* (AD-41) in Norfolk, VA 1984-1987; Reserves in Joplin, MO, 1987-1988; re-enlisted in 1988; served aboard the USS *Missouri* (BB-63) through Desert Storm and until decommissioning on March 31, 1992. Worked in the oil lab and #4 fireroom E.

His memorable experiences include detonation of Iraq's mine; watching the missile attack from the burner front at General Quarters. You can still find his finger prints on the handrail where he braced for the shock. Many other exotic ports completed eight years service.

His medals and awards include National Defense, Southwest Asia w/2 stars, Kuwait Liberation (Saudia Arabia), Good Conduct, Combat Action, Navy Unit Commendation, Battle E Award w/3 E's, Sea Service w/3 stars. Discharged July 1992. After decommissioning of USS *Francis Hammond* (FF-1067).

He married Delinda Auble and has two children. He works as a custom fertilizer and crop sprayer. He is also a volunteer firefighter in Liberty, KS.

CHARLES W. KOLENUT, 6th Div., R Div, Radar, born July 4, 1928, Hawthorne, NJ. He joined the USN June 27, 1946 and assigned to the USS *Missouri* in September 1946. Memorable experiences include meeting President Truman and the death of Cox. Charles was discharged as S1/c April 9, 1948. He is currently a machinist. Married and has one son, one daughter and four grandchildren.

STANFORD L. LADNER, born in Ocean Springs, MS, enlisted USN May 14, 1946. Completed boot training CO-4509, NTC Bainbridge, MD. After boot training, assigned USS *Missouri,* there was about 150 boots assigned to the *Missouri* that same day. He was assigned to the Fire Control Div.

He must thank two fine officers for this assignment, Lt. James Rothermel and Lt. Backzenski. He received some of the best training by the USN from outstanding fire control chiefs. He was advanced to FC2 exactly three years to the day of enlisting. He served four great years aboard the *Missouri.*

After duty aboard the *Missouri,* he served aboard USS *Roosevelt* (CV-42), USS *Magara* (ARVA-6), USS *Vega* (AF-59), USS *Guadalupe* (AO-32), USS *Preston* (DD-795). 1960-1966 assigned to the USNA as instructor of shipyard weapon systems.

Permanent disability retirement as FTC August 1966 after 20 years of active service.

1967 employed by USAF (Civil Service) as electronic instructor at Keesler AFB, Biloxi, MS. Retired from Civil Service 1989 after 21 years service.

Ladner married Mary H. Henley, Gulfport, MS July 5, 1952 and they have two sons, Lee and Kevin. They now reside in Gulfport, MS.

EUGENE W. LAND, born Aug. 9, 1929 in Greenville, SC, enlisted in the USN August 1947, stationed aboard USS *Massey* (DD-778), transferred to USS *Missouri* in April 1948 where he served until discharge in August 1952.

Outstanding memories are of the great liberty ports visited, wonderful shipmates, the long, hard hours working to re-float the *Mo* at Hampton Roads and the exciting days of the Korean War on the first cruise. Of special memory is Christmas Eve, 1950 as they evacuated American forces at Hungnam.

Land received the Korean Service Medal w/3 stars, the UN Service Medal, Navy Occupation Service Medal, National Defense Medal and the Republic of Korea Presidential Unit Citation.

He served in B Div. as an oil king and held rank of BT3 upon discharge. He has two sons, Kelly Eugene and Richard Harrison. He has degrees from Fruitland Baptist Bible Institute, Southeastern Theological Seminary and Luther Rice Seminary. He has pastored Southern Baptist Churches since 1963.

JOHN J. LANE, born Nov. 22, 1934 in Petersburg, MI, enlisted in USN May 26, 1952 and went to Great Lakes for boot camp. Served a short time in Pensacola, FL, before serving on USS *Newport News* (CA-48). Exchanged duty with *Missouri* man to decommission "her". Sailed around through Panama Canal to Bremerton. Quite an experience as "she" had to be hosed down to pass through the canals. Next to last day Feb. 23, 1955 he left and boarded USS *Gardiners Bay,* a sea plane tender. Went into Formosa Strait in defense of Formosa Islands. Flew out, was discharged Nov. 3, 1955 as a coxswain. Received six service medals: Navy Good Conduct, Navy Occupation w/Europe, China Service, National Defense, Korean Service and United Nations.

Lane married Sharon Finney. Their daughter, Julie Lane Kenyon is deceased. Son GMG Monty Lane, career Navy. Lane retired from General Motors, Toledo, OH Sept. 1, 1991 after 30 years.

ROBERT EARL LANE, born June 10, 1928 in Jacksonville, FL, enlisted in USN July 1946 and stationed on the USS *Missouri* from September 1946 to September 1947. After a couple of months in the 5th Div. he was transferred to the K Div. and received his RD-3 in August 1947. He was on board when a five inch starshell round from the cruiser *Little Rock* caused death and damage to the *Missouri* in 1946. Lane later served aboard the USS *Des Moines* (CA-134) as an RD-2. He joined the USNR and was recalled to active duty in August, 1951, serving 15

months on the USS *Chukawan* (AO-100), being discharged as an OS-1 with WWII Victory, European Theatre, National Defense and Good Conduct Medals.

Lane earned a masters degree in chemistry from the University of Florida and worked 29 years for Dow Chemical Company, retiring at Lake Jackson, TX.

PHILIP J. LANOYE, born Dec. 22, 1931 in Buffalo, NY, enlisted in the USN April 28, 1952 and went to boot camp at Great Lakes, IL. He went aboard the *Missouri* in August of 1952, and was on board for two years, I.C.F.C. shop on the 02 level. Transferred to the USS *Des Moines* (CA-134) in early 1954 when the *Mo* went to the west coast to be put out of commission.

He received the United Nations Service Medal, Korean Service Medal, Good Conduct Medal, Navy Occupation Service Medal w/European Clasp.

He was discharged April 3, 1956 as EM3C from Norfolk, VA.

Lanoye was chaplain of the USS Missouri Association for six years and composed the official Missouri Association Prayer. He is married to Rosemary, his wife of 44 years, has a son and daughter and four grandchildren. He is happily retired from the Teamsters.

WALTER J. LASSEN, born Jan. 23, 1918, Marshal, CA, enlisted in USN in 1941. Served on USS *Rodman* (450-DD); USS *Herndon* (DD-638); plankowner on the USS *Missouri* (BB-63). He served as chief gunner's mate with the 40mm quad. His General Quarters station was with the 40mm topside on turret 111. This was rather an active station.

When they started the cruise of the USS *Missouri* from Brooklyn New York, not one of them realized the many military actions they would be involved in. The greatest memory is witnessing the signing of the surrender of the Japanese forces, it will never be forgotten. The huge show of naval forces and hundreds of military planes overhead, gave you the feeling "How dare you attack the USA".

Another great memory, being a guest on the last cruise before final decommissioning of the USS *Missouri* from Hawaii to Long Beach, CA. A great ship and a great crew.

Lassen is in the Navy league. He resides in Maui, HI with wife Dolores and has five children.

ROBERT D. LATHROP, born Aug. 8, 1926 in New Milford, CT, always wanted to join the USN, so he played hookie from high school and went to the next town to join. He went to Sampson boot camp in New York, Newport, RI for firefighting school, then to the USS *Missouri* to put ship into commission June 11, 1944.

From day one his station was on 40mm quad #7 0-5 level starboard side, next to #1 stack. He then had the wheel house going from New York to Turkey. Discharged from Lido Beach, NY, May 14, 1946 as S1C.

He and his wife Jane were married 49 years, September 20, 1997. They have three great children, two boys, David (a state firemarshal); Mark has his own truck business and Jacqueline is a nurse. They also have three granddaughters: Melissa, Michelle and Danielle.

LEO A. LATLIP, born March 16, 1926 in Hallowell, ME, enlisted at age 17 in USN. He attended basic training at Sampson, NY and gunnery training on the USS *Wyoming*. His assignments: S1/c gunstriker, Mount 5, Twin 5-38 cannons, 5th Div. 1944-1946.

He was proud kid from Maine, chosen to serve aboard the most prestigious battleship in the USN, Fleet Commander Adm. Halsey's cabin was located behind Mount 5. Often before muster, they would have a brief friendly visit as he came on deck for fresh air. He was a great man.

He'll always remember how he felt watching the Peace Treaty being signed by General MacArthur and the other dignitaries.

In 1950 he married Meredith Austin of Augusta, ME. They have one daughter, three sons and seven grandchildren. They moved to Washington state in 1967. Latlip retired from Scott Paper in 1989. He enjoys hunting, fishing and traveling.

JAMES P. LAVERY, born May 20, 1932 in Jamaica, NY, entered the USNR March 10, 1951 and active duty Nov. 25, 1953, Brooklyn, NY. He attended Class A Electrical School, NTC Great Lakes, IL.

He served on the USS *Missouri* (BB-63); USS *Northampton* (ECLC-1), USS *Roanoke* (CL-145). His memorable experiences include enjoying those great Mediterranean cruises.

Lavery was discharged in Seattle, WA Dec. 6, 1955 as EMP3 and received the Navy Occupation Medal. He is married and has four children. He retired from McDonnell and Miller, ITT as marketing manager.

DAVID A. LAW, born March 19, 1932 in Lock Haven, PA. Later was in National Champion Drum and Bugle Corps, "The Black Knights". Graduated from Seattle University 1961 with bachelor of commerce and finance. He joined the USN April 21, 1951 and assigned to USS *Missouri* Sept. 15, 1953. He was stationed as telephone talker for captain, executive officer, personnelman.

While at sea the Korean war ended. No action. Ship returned to states when he boarded. He was discharged Feb. 3, 1955 and was awarded the National Defense Medal.

His memorable experiences include when the Secretary of the Navy was dipped into the ocean from boatswain's chair on ship to ship transfer, when men pulling high line, tripped and fell and line went limp.

He married in 1955 and is presently divorced and retired. He has three children.

MICHAEL C. LAWTON, born Dec. 6, 1964 in Albuquerque, NM, enlisted in USN September 1985. Stationed on USS *San Bernadino* (LST-1189) forward deployed in Sasebo, Japan. Served aboard USS *Missouri* (BB-63) from June 1989 until discharged September 1991.

Memorable experiences include being named Sailor of the Quarter (3rd Q, 1990); meeting Adm. Larson (CINCPACFLT) in Pearl Harbor on the way to Desert Storm; shooting of numerous Tomahawk Missiles and 900 plus 16 inch rounds at Iraq and Kuwait, even a Silkworm missile being shot at them. After all that the most frequently asked question he received when they docked in Australia was "Is this the ship Cher made her video on?" When your famous, your famous!

He achieved the rank was ET1 (SW) and was awarded Navy Achievement Medal w/Gold Star, Navy Unit Commendation, Combat Action, National Defense, SWA Service w/stars, Kuwait Liberation, Surface Warfare Specialist Insignia, Navy "E" w/4 E's, Sea Service w/2 stars, and Good Conduct Medal.

Lawton is married to the former Penny Tolliver and they have one son, Chad. He received an AAS in electronics and is now employed at Sandia National Laboratories in Albuquerque, NM.

FRANK P. LECHNAR, born May 27, 1926 Uniontown, PA, was inducted into the USN March 17, 1944 in Sampson, NY, and assigned to the USS *Missouri* April 27, 1944, plankowner.

He served in 3rd Div. Center gun, Turret 3. His memorable experiences include the surrender ceremony; supporting the landings on Iwo Jima and Okinawa; bombardment of the home islands.

He was awarded the American Campaign, Asiatic Victory w/3 Battle Stars, WWII Victory Medals and Asia Occupation.

Lechnar was discharged May 12, 1946 in Bainbridge, MD. He married Virginia Veshinfsky, April 30, 1954 and they have two sons and a daughter.

WARREN S. LEE, born Sept. 22, 1928 in Mount Vernon, NY, enlisted in USN Oct. 5, 1948. Attended boot camp then Fire Controlmen School at Great Lakes, IL.

Boarded the *Missouri* May 1949 and was assigned to the FM Div. His GQ and working station was the main battery plotting room forward from where he fired the 16" guns.

Lee was awarded the Navy Commendation Medal for Meritorious Service when *Missouri* was engaged in bombarding enemy targets in North Korea from September 1950 to March 1951. He also received the Good Conduct, China Service, Navy Occupation, National Defense, United Nations and Korean Service Medals.

He served 39 months on the *Missouri*, attaining the rate of FT2c. He was discharged in August 1952. Lee retired in 1988 after being with IBM for 36 years.

Lee is married to Nann Wagner and they have three sons: Douglas, Steven, Neil and one daughter, Debra Preiser and nine grandchildren.

CLAYTON W. LEIST, born March 12, 1918 in Cincinnati, OH, was inducted into the USN Oct. 23, 1943 and assigned to USS *Missouri* May 1944 to December 1945, Turret #2 E Div. and sound power telephone shop.

His memorable experiences include when kamikaze hit under fantail and shelling of Hokkaido. While at sea he participated in the Okinawa campaign.

He was awarded the Victory Ribbon, Asiatic-Pacific w/three Stars and American Theatre.

Leist is married to Jean Locke and they have two sons, one daughter and two stepdaughters and a stepson. He is retired from Formica Corp. after 38 years of service.

RUSS LESTER, born in Kalispell, MT on Dec. 25, 1912, grew up on Huntington, NY, was drafted at the ripe old age of 30 in late 1943. Because he was older than many on the ship he was often referred to as Pops.

He joined the big *Mo* early in its career and served as an electricians mate third class through it's Pacific tour, until he left the ship in New York in October of 1945.

Lester and his wife Ann retired from the New York Telephone Company in 1975 with a combined service time of over 65 years.

They toured the US in a travel trailer for many years after their retirement and have been regular attendees at the *Missouri* reunions.

He and Ann are living at "A Villa Rosa" in Palm Harbor FL. They have three children, 10 grandchildren and eight great-grandchildren.

GERALD A. LINDSTROM, born Nov. 10, 1919 in Austin, MN, enlisted in USN Feb. 13, 1940, served

aboard the USS *Wichita* on the neutrality patrols, Russian convoys, invasion of North Africa, the Guadalcanal, Attu, and Kiska campaigns. March 18, 1944 assigned to USS *Missouri* precommissioning detail as MM1c responsible for air conditioning and refrigeration. Later at Ulithi was made Master at Arms for A Div.

He noticed Kennealy sketching cartoons and asked if he could put ideas into cartoons, such as, not wasting water. Being a hit the editor said no need to be instructional so tell Kennealy use his imagination. Thus the Boswell was created that gave the crew a hearty laugh every morning.

He was awarded the American Defense w/ Bronze A, American Theatre w/Star, European-African w/4 stars, Asiatic-Pacific w/5 stars, Victory and Good Conduct.

He is married to Patricia Suerth and they have four daughters, two sons, four grandchildren and two great-granddaughters. After 34 years in construction management retired from The Zia Co., Los Alamos, NM in 1987.

LELAND W. LINDSTROM, born Jan. 23, 1928 in Crystal Beach, Ontario Canada, entered the USN Dec. 3, 1947 and assigned to the *Missouri* March 1948 as boiler technician. His memorable experiences include running aground. While at sea he participated in bombarding Korea. He was discharged Dec. 4, 1951 as BT3.

He is married and has four children. Now retired and living in the Allegheny Mountains.

ANDY LINETTE, born March 24, 1927 in Louisville, KY, enlisted March 24, 1945 and served aboard the USS *Tomahawk*, operating from Ulithi in the Caroline Islands and participated in fleet operations for the occupation of Japan. He was assigned to the USS *Missouri* in February 1946 as yeoman third class in the EX Div.

Mehmet Munir Ertegun, the Turkish Ambassador to the US, had died on Nov. 11, 1944 and his body rested in Arlington National Cemetery until March 22, 1946 when the body was placed aboard the USS *Missouri* for the voyage to Istanbul. This was the beginning of a memorable trip to the Mediterranean. Ports of call included Gibraltar, Istanbul, Athens, Naples, Algiers and Tangiers. He was privileged to have an audience with Pope Pius XII with some of his shipmates, which was arranged by Chaplain Paul O'Connor. Today, Andy and his wife, Catherine live in Cape Coral, FL.

HAROLD J. LOEFFLER, born Feb. 25, 1926, Newark, NJ, enlisted Jan. 27, 1944. He graduated high school and attended boot camp and Hospital Corps School Great Lakes, Brooklyn (NY) and St. Albans (NY) Naval Hospitals.

Awed by the size and armament of the USS *Missouri* when he boarded in August 1944 as she laid at anchor in Gravesend Bay, NY. Left on shakedown cruise shortly after.

His memorable experiences includes: Panama Canal; entering Pearl Harbor; shelling of Japan and Okinawa; seeing kamikaze hit ship; surrender ceremony; trips to Turkey and Mediterranean ports; audience with Pope Pius 12th.

Loeffler was discharged as PhM3 with American Theatre, Asiatic-Pacific w/3 stars and Victory Medals on May 13, 1945, Lido Beach, NY.

He married Julia Rundquist and they have four daughters, (one deceased), one son, nine grandchildren. Retired from Asbury Park (NJ) Press Advertising Department. He resides in Toms River, NJ.

MICHAEL J. LOGUE, born Dec. 30, 1950 in Montebello, CA, joined the USN June 11, 1977 and was assigned to the USS *Missouri* August 1989-August 1991, as senior medical officer, medical department head. He achieved the rank of captain.

While at sea he participated in the coordination of all medical support during Operation Desert Shield/ Desert Storm, including periods of 16 inch gun shore bombardment, cruise missile launches, scud missiles being fired at the ship, and a reported poison gas attack.

His memorable experiences include having the honor of visiting with medical department shipmates from WWII and Korea in their spaces; bandaging hundreds of crewmembers fingers cut in their attempt to remove gas mask filters from their canisters at the start of "this is not a drill" gas attack during the Persian Gulf War; being the CO's running-mate on liberty.

Logue is presently still active and has received three Meritorious Service Medals, Combat Action Ribbon, two Meritorious Unit Commendations and two Battle E.

He is married to Christine Fischbach and they have two daughters, Sarah and Heather. He is director of Managed Health Care, Naval Hospital Bremerton.

ANTHONY J. LOMBARD, born Dec. 4, 1925 in Waterbury, CT, was inducted into the USN March 17, 1944 and assigned to the USS *Missouri* May 1944.

His stations include: FM Div. radar operator, New Haven, CT; Sampson, NY; Newport, RI; USS *Missouri*; Brooklyn, NY; Lido Beach, NY.

His memorable experiences include: commissioning June 11, 1944; on *Mo* in the Pacific February 1944, appendix operation; surrender signing Sept. 2, 1945.

While at sea he participated in Okinawa and Iwo Jima Campaigns; Air support for aircraft carriers.

He was awarded the American Theatre Medal, Asiatic-Pacific w/3 stars and the Victory Medal and discharged May 13, 1946 as S1c.

He is married to Phyllis Ann Morris and they have two sons, two daughters and one grandson. He is now retired from Lombard Bros. Inc. (common motor carrier).

MAYNARD E. LOY SR, born May 27, 1932 in Falls Church, VA, enlisted into the USN, July 26, 1951 and went through boot camp at the US Naval

Training Center in Great Lakes, IL. He reported aboard the USS *Missouri*, Nov. 22, 1951, 4th Div. gun gang, and spent a three year tour of duty aboard her. There were a lot of memories but his most lived, was while they were in Korea 1952-1953. There were some good times and there were some bad times.

He remembers they were in Wonson Harbor one morning on routine shore bombardment when they received counter battery. All nine of their 16" guns (turret 1, turret 2, turret 3) trained to their port side and fired a nine gun salvo or broadside. This operation is very rare for a battleship as it puts so much strain on the ships superstructure. However they got the job done by pushing the enemy back so their ground forces could advance.

There was a most saddened experience when they lost their helicopter pilot, Ensign Robert L. Mayhew, while on a spotting mission, Dec. 22, 1952. On their way home to the states in Sasebo Harbor, Japan on the morning of March 26, 1953 they lost their captain, "Warner E. Edsall" of a heart attack, while on the bridge making preparations to anchor.

The *Missouri* was the last battleship to fire her main battery 16" guns in Korea, before the truce was signed. She was his world at sea, his home in port, a fighting lady, with a heavy punch. He will always remember her.

Loy left the *Mo*, in 1954 before she left for Bremerton, WA to be decommissioned and reported aboard the heavy cruiser the USS *Des Moines* (CA-134). He finished up his enlistment aboard the battleship USS *Mississippi* (EAG-128) and was discharged from active duty July 26, 1955, rate gunners mate third class.

He received the Korean Service Medal (w/3 Battle Stars), United Nations, China Service, National Defense, Presidential Unit Citation (awarded to ship by Syngman Rhee, President Republic of Korea), Navy Occupation and Good Conduct Medal.

He and his wife Nancy have two sons, Wayne and Maynard Jr., and two daughters, Marion and Michelle, a daughter-in-law, Martha, Wayne's wife, a son-in-law, Mike, Michelle's husband, four granddaughters: Melissa, Virginia Lee, Katherine and Christine and two grandsons, Kenneth and James.

He has his own appliance repair service in Falls Church and plans to semi-retire some time in the near future, but the ole saying is" "Keep working and keep young."

DAVID W. LUCAS, born Dec. 30, 1927 in Pittsburgh, PA, enlisted in the USN December 1945 and was assigned to the USS *Missouri* January 1946, after completing boot camp at Norfolk, VA.

His most memorable cruises include: Turkey, getting hit with friendly fire from USS *Little Rock;* Rio de Janeiro; cruise with President Truman, Mrs. Truman and daughter Margaret; Trinidad and crossing the Arctic Circle.

He was discharged Dec. 4, 1947 in Brooklyn, NY and went to work at the Washington Navy Yard, January 1948 and worked there for 35 years. Retired

January 1983 and worked for nine years as a bailiff in Prince Georges County, MD

He is married to Jean and has five stepsons and one stepdaughter, six grandsons and four granddaughters.

Lucas received the American and European Theatre Medals and the Victory Medal. He plays golf two or three times a week and enjoys working in the yard with Jean.

JOHN H. LYNCH, born Feb. 18, 1924 in New Britain, CT, joined the USN May 21, 1943 and was assigned to the USS *Missouri* upon commission.

His memorable experiences include signing of the surrender on board.

He was discharged March 12, 1946 as S1c and was awarded the American Theatre Medal, Victory Medal, Asiatic-Pacific Theatre Medal w/3 stars.

Lynch died Jan. 30, 1995. He was retired from the US Post Office after 40 years. He was married for 43 years and had two sons, Christopher M. Lynch and Michael J. Lynch; one daughter, Ann Marie and grandchildren, Patrick Lynch and Alicia Anop.

ROBERT G. MACKEY, LT(sg), born Oct. 3, 1920, Chicago, IL and grew up in Chicago's West Side. He attended Austin High School and played football on the same team with Chicago prep legend, Bill DeCorrevont. As an undergraduate at Northwestern, Mackey played baseball and was highly rated in the back stroke.

Mackey was recruited into the USN Supply Corps and received his commission in 1942, reporting for active duty a year later. After serving in various US ports during the first part of his military career, Mackey was ordered to sea and joined the crew of the USS *Missouri* in late October 1944. Shortly thereafter, the Missouri began its voyage to Hawaii then on to the Pacific. From January to March 1945, it was underway as part of various task groups which joined on air strikes and bombardments against Tokyo, Okinawa and Kyushu, as well as supporting the invasion of Iwo Jima. From late March to May they provided air support for the Okinawa Campaign and joined in air strikes against Kyushu and other task units for the bombardments of Hokkaido and Honshu. Lt. Mackey was the disbursing officer and responsible for GQ, the Coding Ward, and the Stewards's Division.

After the surrender, Mackey managed the Missouri's baseball team, which defeated the Caribbean Champions under his leadership. In March 1946, the *Missouri* transported the remains of Turkish ambassador, Munir Ertegun, from Washington, DC to Istanbul.

Returned to civilian life in 1946 and began law school at Loyola University. Went into private practice but continued to serve in the USNR for more than 20 years, soon transferring from the Supply Corps to the Judge Advocate General Office. A highlight of his career occurred when he was in Washington to be sworn in and then Vice President Nixon invited them to his office for a visit.

As a judge, Mackey sat on the Circuit Court of Cook County for 17 years. He is still an avid swimmer and can often be seen swimming at the club.

MORGAN N. (JIM) MADDEN, born March 31, 1927 in Plain Dealing, LA, quit high school at the age of 16 to join the USN in 1943 as WWII raged on.

He was first assigned to the USS *Walker*, but in 1944 he was sent aboard the USS *Missouri*, where he

served until 1947. He saw action in Okinawa, Iwo Jima and other areas of the Pacific. He has always been proud of the fact that he was aboard "*Big Mo*" when the peace treaty was signed.

He served the balance of the war aboard the USS *Little Rock* (CL-92) and was discharged in September 1949.

Before retiring he worked as a rigger, ironworker out of Local 27, in Salt Lake City, UT. He always says he got his education in the USN.

He married Norma Helquist in 1953, and has five children, 20 grandchildren and two wonderful great-grandchildren.

JOHN (JACK) C. MALEY, born July 19, 1926 in Chicago, IL, enlisted in the USN September, 1943. Stations: Great Lakes, IL; Norfolk, VA; Destroyer School; Newport, RI; Gunnery School; Brooklyn Navy Yard assigned to USS *Missouri* when commissioned. Member of the 10th Div., 40mm until discharged March 1946.

He remembers when the 3rd Fleet occupied the Yokosuka Naval Base and being part of the occupational force that was sent over to the base.

He was awarded the American Area, Asiatic-Pacific w/3 stars, WWII Victory Ribbon, Third Fleet w/1 star and Philippine Liberation.

Maley is married to Muriel Sutfin and they have two daughters and four grandchildren. He retired from city of Chicago Police Department after 36 years.

PERCY WARNER MALLISON, born Oct. 22, 1920, Great Lakes Naval Station. He joined the NROTC in 1941 and was assigned to the USS *Missouri* Feb. 24, 1944, 8th Div., gun officer.

While at sea his battle station was Sky 1-A, Starboard forward 2-40mm quads and 1-5"38 dual on director.

His memorable experiences includes a kamikaze burning under his 5"38 mount; ordered to flight training and transferred at sea.

Mallison was discharged in 1946 as LTJG USNR plankowner. He is married to Mary and they have four children and six grandchildren.

FRANK D. MANCINI, born in Lawrence, MA Oct. 5, 1920, joined the USN Dec. 17, 1942 and assigned to the USS *Missouri* 1944, as corpsman PhM2c.

While at sea he participated in the Asiatic-Pacific.

His memorable experiences: as a physical therapist attended many dignitaries, including Admiral Wm. Halsey.

He was discharged November 1946 and was awarded the American Theatre Medal, Asiatic-Pacific w/3 stars and the WWII Victory Medal.

Mancini is married and has three children and five grandchildren. He has been a chiropractic physician past 48 years.

GEORGE EDWARD MANN, born May 8, 1930, Millard Filmore Hospital, Buffalo, NY, joined the USN Jan. 15, 1951 and assigned to the USS *Missouri* August 1952, exec's office.

While at sea he participated in many gun strikes, Korean Theatre October 1952-April 1953. Operating as 7th Fleet Flag (Adm. Jocko Clark). He was discharged Dec. 2, 1954 as PN3.

His memorable experience: loss of spotter helicopter Dec. 23, 1952 with all aboard (3) during combat gunnery. Only two bodies recovered; he knew all three men.

He was married for 43 years to Sara Beatty Mann, and they had five daughters, ten grandchildren. Sara died May 3, 1996. He is a CEO and majority stockholder, ESS-Kay Yards, Inc. a Marine facility in Brewerton, NY.

JOSEPH P. MANNING, born Dec. 6, 1924 in Boston, entered into service April 10, 1952 and assigned to the USS *Missouri* June 1952, 7th Div. as deckhand.

While at sea he participated in the Korean campaign and was awarded the UN Medal, Korean Medal w/2 stars and China Good Conduct. He was discharged Dec. 6, 1955.

HENRY C. (HANK) MARCHESE, born April 29, 1948 in New York City, enlisted Aug. 28, 1967. Commissioned as ensign (LDO) April 1, 1980. Retired July 1, 1992. Stations: RTC and Fire Control School, Great Lakes, USS *Tripoli* (LPH-10); USS *Paul Revere* (LPA-248); Purdue University, USS *Richard L. Page* (FFG-5); Guided Missile School, Dam Neck, VA; USS *Conyngham* (DDG-17); Atlantic Fleet Weapons Training Facility, Roosevelt Roads, Puerto Rico; Advanced Undersea Weapons School, SSC, Orlando, FL; USS *Missouri* (BB-63) (until decommissioning).

He received a BS in computer science. Fire control officer. Among the things remembered that are too numerous to mention are great shipmates, and his retirement ceremony on the surrender deck just prior to BB-63 decommissioning. Received the following: Surface Warfare Breast Insignia; Medals: Navy Commendation (4-1 w/Combat V), Navy Achievement (2), National Defense Service (2), Good Conduct (3), Republic of Vietnam, Service Campaign (7), and Presidential Unit Citation w/Oak Leaf of Valor, Southwest Asia Services (2), Kuwait Liberation (Kuwait), Kuwait Liberation (Saudia Arabia); Ribbons: Combat Action, Navy Unit Commendation (2), Battle E (4), Sea Service Deployment (4) Overseas Service (4), Pistol Marksmanship, Rifle Marksmanship.

Marchese is married to Theresa Gauden and they have two sons. After retirement from the USN he attended the University of Central Florida where he obtained his degree in Computer Science. He is a software engineer and has recently been promoted to director of operations, and research and development for Jenoptik INFAB, Inc. (USA).

ERNEST T. MATHEWS, born Sept. 22, 1926 in Turtle Lake, ND, entered the USMC, Jan. 20, 1945

and was assigned to the USS *Missouri,* May 20, 1945, serving 20 and 40mm, later captain's orderly, 1946.

His memorable experience includes surrender ceremonies, Panama Canal and Turkey.

While at sea he participated in the end of Okinawa and bombarding Japan.

He was awarded the Asiatic-Pacific w/2 stars, American Theatre and Victory Medal. He left the *Missouri* in Portland, ME, July 31, 1946.

Matthews married Jacqueline Nunn July 8, 1948 and they have one son and two daughters. He retired as painter after 12 years. His hobbies include antique cars, 36 Ford pickup, 39 Coupe, 65 Plymouth Convertible.

CHAS. W. H. MATTHAEI, born May 6, 1920 in Tacoma, WA and attended Tacoma Public Schools. Entered the University of Washington in 1938, majored in chemical engineering, also enrolled in Naval ROTC. Received a commission as ensign DE-V(G).

Graduated University of Washington in chemical engineering June 1943. Received orders to Ordinance and Gunnery School Navy Yard, Washington, DC, in December 1943. After completing antiaircraft courses in fire control and gunnery, assigned to the pre-commissioned detail of BB-63 at NTS Newport, RI, and participated in training the deck crew.

Boarded the *Missouri* in the Brooklyn Navy Yard with the *Newport* crew on June 10, 1944, for commissioning the following day. Assigned to the FM Div. After the shakedown cruise, the ship went through the Panama Canal to San Francisco for modifications.

While the ship was in San Francisco, attended machine gun school at Point Montara, CA, and then assigned to the 8th Div. (machine guns).

Participated in defending the ship against many air attacks until the war ended. With the war over, gunnery was no longer interesting, and Matthaei requested and received a transfer to engineering. Matthaei remained in engineering, M Div., until he separated from active duty in July 1946. Resigned his commission with the rank of lieutenant in 1958.

He married Helen Leon, a WAVE officer, and they have three sons and seven grandchildren. He is still active in business, where he is chairman of Roman Meal Company, Tacoma, WA.

ALLEN F. MAXFIELD, born Oct. 11, 1923 in Brooklyn, NY, joined the USN August 1941 and assigned to the USS *Missouri* Sept. 2, 1946, as A Div. officer.

While at sea he served three years on *Mighty Mo.*

His memorable experiences include bringing President Truman and wife, and daughter from Rio to the states.

He was discharged June 1949. Now retired from U.S. Testing Co. after 39 years. He is married and has three children and ten grandchildren.

JAMES H. MCCLELLAN, born Jan. 21, 1923, Zolfo Springs, FL, enlisted in USN Nov. 5, 1952, Tampa, FL. He received training at Norfolk, VA. Assigned to the USS *Oasterhouse* plankowner to Europe, Africa to Guadalcanal, Bougainville, Florida Island, Solomon Island and New Guinea. Returned to San Francisco, CA to Newport, RI to Brooklyn, NY and

assigned to USS *Missouri* plankowner until transferred to USS *New Jersey* 1947.

His memorable experiences: lost 40mm gun mount during Iwo Jima, Okinawa, operation and typhoon. To one other gun mount that was on the receiving in of the wing of a kamikaze nearly destroyed this gunmount; to see the Japanese surrender Tokyo Bay; two friends were killed on the USS *Missouri;* the Arctic Circle special assignments. He was a barber.

He received the American Campaign Medal, EAME Campaign Medal, Asiatic-Pacific Campaign w/6 stars, Occupation Service, WWII Victory Medal, Good Conduct.

He is a veteran of Korean World. McClellan is retired and happily married.

ROBERT D. MCCRINDLE, Return from Istanbul: while on a "speed run" from Gibraltar to Norfolk, VA assigned task of preparing a company of seamen from the 4th Div. to take part in a parade in Boston to celebrate VJ Day 1946.

Attempting to march on the fantail while underway was something. Unsuccessfully sought permission to "blouse" the bellbottoms over the leggings to forgo "bootcamp" appearance. Arriving at staging area for line of march discovered their place was immediately in front of the 82nd Abn. Div. replete with scarves and stainless steel helmets. However, the *Missouri* won the day with much cheering and its picture on front page of newspapers the following day.

Summer in Portland, ME: Selected to wear experimental new USN uniform. Overseas cap, Eisenhower jacket, trousers, shirt and tie. Wore during summer in Maine. Subsequent inspection was asked if he liked it. "No, Sir, too much upkeep" he replied. Back to bellbottoms!

Norfolk, VA: his last Christmas aboard ship, 4th Div. won first prize decorating living compartment. Transformed into a living room with fireplace (ammo boxers) with stockings hung and all the trimmings. He believes it helped cheer up some homesick seamen on their first holiday away from home.

FRANK MELLO JR., born Oct. 21, 1929 in New Bedford, MA, joined the USN Aug. 4, 1948, and was assigned to the USS *Missouri* Jan. 13, 1949 as commissary man.

While at sea he practiced shooting 40mm guns and aeroplane.

His memorable experiences: going to Portsmouth, England; Guantanamo Bay, Trinidad.

He was discharged Aug. 5, 1949 as CSSM. He spent five more years in the Reserves and was discharged Sept. 15, 1954.

Mello is retired after 35 years as a civil engineer. He has two daughters and one son. His son became a doctor and is employed at St. Elizabeths in Boston and has three children.

JOSEPH L. MELOIA, born Oct. 7, 1931 in Newark, NJ, joined the USN Feb. 14, 1951 and was assigned to the USS *Missouri* June 1951, as storekeeper.

While at sea he participated in patrolling and shelling the coast of Korea.

Serving aboard the USS *Missouri* was the best experience.

He was discharged Feb. 14, 1955 and awarded the United Nations Medal, Korean Service Medal and China Sea Medal.

Meloia is married to Rose Marie and has two daughters, Robin and Dawn. He is supervisor of classic tile in Elizabeth, NJ.

BRUCE C. MENELEY, born March 20, 1957 in Bridgeport, CA, joined the USN Aug. 4, 1975 and was assigned to the USS *Missouri* Dec. 15, 1989-March

31, 1992. He was the last battle-ship medical officer in the USN and decommissioned the ship.

He served as medical officer and received the rank of LCDR. He participated in Desert Shield/ Desert Storm.

He attained the designation as a surface warfare officer, and received the Navy Commendation Medal, Combat Action Ribbon, Navy Unit Commendation, Combat Efficiency Ward w/2 awards, National Defense Medal, SouthWest Asia Service Medal w/2 Bronze Stars, Sea Service Deployment w/2 Awards, Kuwait Liberation Medal (Saudi Arabia), Government of Kuwait Liberation Medal.

Meneley is currently staff physician, Emergency Department, Naval Medical Center San Diego. He is married and living with his wife Karen in Escondido, CA.

MICHAEL METTLER, first saw the *Missouri* in December of 1943 after returning to the States from a tour of the Pacific sponsored by the USMC. Among the islands toured in 1942 and 1943 were Guadalcanal, Efate, Espirito Santos and New Caledonia. The last islands were to visit various Navy hospitals and then home.

Stood guard duty on many a bitter cold and windy night that January and February 1944, at Pier G, where the *Missouri* was being completed. So bundled up that he could not have taken the weapon off his shoulder to use if anything suspicious had occurred.

Then "Shanghaied" on the *Missouri* a couple days after the commissioning. Reported aboard and was immediately informed that he would have to divest himself of any and all ribbons, whether or not he was entitled to them. This led to a minor problem which was later solved.

The ship then moved to Hoboken for degaussing and they played some ball on the pier. Thence to Norfolk and liberty and on to Trinidad. During the return trip they had an exercise with "pocket battleship" which he believes was the USS *Alaska* and then homeward.

They became involved in a hurricane off Cape Hatteras in September. He remembers clearly being ordered at one time to climb up a smoke stack with another crewman, for "submarine watch" ridiculous! They strapped themselves to the stanchions of the small "walkway" that was at the top of the stack. When they were relieved, their foul weather gear was absolutely white from the sea salt.

As his health record was stamped in bright red ink to paraphrase - "No duty outside the continental limits of the U.S." he felt it was time to leave the ship. This was later done in NYC. So he can say he served before and after the commissioning for about one year.

He recently retired from the business world, for the third time, and taking life a little easy. He may seek work, but who knows.

LEWIS M. MILLER, born Nov. 6, 1930 in Vandergrift, PA, enlisted USN Dec. 3, 1947. Reported aboard *Mo* March 13, 1948, as SA, assigned to 4th Div., as gunner's mate striker in Mount 4.

He really enjoyed the *Mo,* being 17 having the opportunity to see the world, and learning to grow in the best USN ever. He didn't know that then, but he does now. He left the *Mo* as GM2 on March 13, 1954.

He went on to serve on USS *Worcester,* USS

Bennington, USS *Coral Sea,* USS *Lake Champlain,* USS *Kitty Hawk,* USS *Canberra,* USS *Farragut.* He retired as GMMC on March 13, 1967 at Naval Weapons Station, Chosin, SC.

He married Betty Eshelman and they had three sons: Todd, Lewis, Douglas, and daughter Susan and seven grandson. They have been married 45 years.

He retired from PMFLANT, Chosin, SC as missile inspector, with 25 years on Oct. 1, 1992 and is now living in Sebring, FL.

HOLLIE M. MIMS, born July 19, 1919 in Ozark, AL, enlisted Oct. 4, 1939. During his career he served aboard the following: USS *Vincennes,* escort duty in the North Atlantic; harbor defense unit, Little Creek, VA; Communications Unit, New Zealand; PCE-891; USS *Albany;* USS *Missouri* 1946-1950; Flag Staff, commander UN Blockade and Escort Force; US Navy Recruiter, Tallahasse, FL; LST-722; USS *Aucilla,* home ported Barcelona, Spain; Harbor Defense Unit, Coco Solo, Canal zone; USN recruiter, Daytona Beach, FL. Retired July 31, 1962.

During any Navy career there are many unforgettable events. One of which was when the *Might Mo* brought President Harry Truman and his family back from a conference in Rio. Also, being aboard when the ship ran aground in Hampton Bay.

Another when aboard the USS *Vincennes* bringing back millions of dollars of gold from South Africa for safekeeping coming via Recife, Brazil.

He married Edna Lorraine Shinkle, Brooklyn, NY Jan. 31, 1948. They had one son, David Allen Mims, his wife Pamela, grandsons, David Jr. and Billy. Spent time gardening and volunteering as a member of many organizations. He died May 13, 1997, Daytona Beach, FL.

After retirement he worked on the Apollo Project at Cape Kennedy with General Electric Company. Also on special projects in Turkey and the Philippines.

HENRY C. MOITY, born Nov. 2, 1925 in New Iberia, LA, joined the USN January 1943 and served on USS *Ranger* (CV-4) in Atlantic 1943-1944. Served USS *Missouri* 1945-1946.

His most memorable experience: he was member of boxing team, middleweight on both ships. He received one Battle Star on USS *Ranger* and three on the USS *Missouri.*

Moity was discharged as S1/c in 1946. He joined the US Police Department (retired) and joined a law firm, Phelps, Dunbar as chief investigator on maritime cases, retiring in 1991.

Moity has a son, Michael, daughters, Denise, Susan and Dayna, four grandsons, and three grand-daughters.

JAMES O. MONKHOUSE, born to JL and Bessie Arilla Monkhouse in Shreveport, LA, Dec. 23, 1915, entered the USN April 1944. He served aboard the USS *Missouri* from December 1944 until December 1945. Discharged on Dec. 18, 1945 as

firecontrolman third class. Using the GI Bill, he entered college in 1948, receiving a degree in commerce in 1952.

He entered the USN married to Bessie Boykin. Sixty years later they have three daughters, nine grandchildren, and five great-grandchildren.

After 31 years with Libbey-Owens-Ford Company, he retired to Louisiana, later moving to Arkansas where he has been for six years, churching, fishing, golfing, and hunting with his grandsons and son-in-law. He still maintains correspondence with two shipmates, Lindsey Edwards and Al Besemer.

Although many friends, suicide plane, near miss in the heat of battle, a storm makes it difficult choosing greatest memory, it has to be the ceremony of signing the treaty.

JOHN R. MORANO, born May 1, 1930 in Brooklyn, NY, enlisted in the service January 1951. Attended boot camp in Newport, RI, Co. 151; Radio School, Norfolk, VA; 30 days TAD aboard USS *Midway.* Assigned to the flag division of COMBATCRULANT where he served aboard the USS *Missouri,* USS *Albany* and USS *Des Moines.*

Adm. Holloway was in charge. Later Adm. Wooldridge took over. Serving in a flag division for an admiral is number one duty.

Flag Divisions were made up of radiomen, cryptographers, teletype, navigation, signalman, photographers, boatswain mates (Adms. Barge). They just did their work and because they were flag, they were excused from all other shipboard duties.

When the *Wisconsin* had to go back to the Far East Morano transferred to ships company, eventually he then transferred to the flag division of Commander 7th Fleet, under Adm. Clark, later Adm. Pride. He served aboard the USS *Rochester,* USS *St. Paul* and the USS *Helena.*

He made two midshipman cruises to the *Missouri* (1952 Norway and England, 1953 to Rio de Janeiro). Between the *Missouri, Albany* and *Des Moines* they hit ports in the Caribbean Sea and a mess of ports stateside. He crossed the equator, the Arctic Circle, International Dateline. He was honorably discharged January 1955 as RM3.

Morano belongs to the Missouri Association, Wisconsin Association, ABA. At a reunion in St. Louis, he met his division officer and flag crypto officer from the *Wisconsin* after 42 years. They now keep in touch.

He earned the Good Conduct Ribbon, Korean, UN Ribbon, National Defense and China Service Ribbon. He has been to many ports in the Far East from Japan, Hong Kong, Taiwan, Philippine Islands, Korea. Because they were flag they hit many ports in Japan.

Morano is now retired and lives in Roslyn, NY and Flushing Queens, NY.

ALFRED C. MORGAN, born Aug. 25, 1930 in Taylorsville, GA, entered the USN on May 10, 1948. Attended boot camp in San Diego and served aboard the USS *Missouri* starting in September 1948. After serving approximately two years in the #2 Fire Room he transferred to the oil gang where he remained until discharged in May 1952.

His memorable experiences include: meeting President Truman aboard the USS *Missouri;* running aground in the Chesapeake Bay, VA; visiting numerous foreign ports on midshipman and reserve cruises; and as Goodwill Ambassadors. The significant memories of all are the life long friends he made while serving on the *Missouri*. He remains in contact with many of them today.

He married Shirley Davis and has two sons, two daughters and nine grandchildren. Retired from Bowman Transportation and is an active member in church, Masons, city and county activities, historical vehicle associations, and keeping up with the grandkids.

WILLIAM A. MORIN, born April 13, 1926 in New Britain, CT, joined the USN March 14, 1944 and was assigned to the USS *Missouri* June 11, 1944, stationed on gun mount #3.

While at sea he participated in Iwo Jima, Okinawa, bombarding Tokyo and being an honor guard to the surrender ceremony.

His memorable experiences include being locked in gun mount #3 when it was on fire, after the suicide plane hit the ship and riding out a typhoon.

He was discharged Jan. 17, 1946 as S1c and was awarded the WWII Victory Medal, Asiatic-Pacific Medal w/3 stars and the American Area Medal.

Morin has been married 38 years to Patricia Greenlaw and they have a son, Robert Morin, daughters, Caryn Dias, Susan and Nancy Morin, seven grandchildren and one great-grandchild. He is now retired from the Fall River Housing Authority.

WILLIAM D. MORTON SR., born Feb. 10, 1932 in Quincy, MA, joined the USN May 8, 1950 and was assigned to the USS *Missouri* August 1950 as boatswain mate.

While at sea he served with the 40mm gun crew as coxswain on all types of small boats and landing craft.

He was discharged April 1954 as BM2c and received the Korean Service Medal w/5 stars, Korean Presidential Unit Citation, United Nations Medal, China Service, Occupation, National Defense and the Good Conduct. Morton was married to the late Mildred A. Morton and they had three children and five grandchildren. He is a retired paper hanger.

WILLIAM FLOYD MOWDER, born April 5, 1933 in Washington, NJ, was inducted into the USN May 23, 1950, assigned to Pier #7 August 1950 in Norfolk, VA, fireroom #2, lower level B Div.

While at sea he participated in all campaigns the *Missouri* took part in from 1950-1954 and Korea. He was awarded the Navy Occupation Service Medal, China Service Medal, National Defense Service Medal, United Nations Medal and the Korean Service Medal w/2 stars.

March 22, 1952 he had just reported to Fireroom #2 to relieve the 4 a.m. 8 a.m watch for early chow. He was looking forward to a good breakfast himself after they ate. Just as they returned from chow, general quarters sounded. They were a couple miles off the coast of Wonsan and was fired on, but they were not hit. It was the closest enemy shelling in two Korean combat tours.

Despite the Red counter fire, the third attack they received, they destroyed two heavy gun emplacements, closed three caves, and many secondary explosions on their 11th bombardment of the Red coastal city 80 miles north of the 38th Parallel.

After shelling Wonsan, they steamed south to Kojo Island and destroyed three more gun positions, six supply buildings and badly damaged four others. He did get to eat later on, but he can't remember when it was.

He was discharged Jan. 25, 1954. His father, Samuel L. Mowder passed away in 1979 and his mother, Ruth C. Mowder is still going strong 82 years young. His wife, Doris J. passed away in 1985. He remarried November 1990 to Laura Darlington and has one son, Wm. Dale, two daughters, Lou Ann Pursell and Susan Lynn Ulmer, one stepson, Gerald and two stepdaughters, Laura and Sandra. He owned and operated a service station and has just sold out to his son and is retired.

ROBERT A. MOYER, born Sept. 26, 1922 in Reading, PA, enlisted in the USN May 20, 1942 and was assigned to the USS *Missouri* October 1948 to May 1953. He served in B Div., #2 Fireroom and was aboard from 1948-1953.

His memorable experiences include the many visitors and dignitaries to see surrender plaque; hurricane 1950 off Hatteras while enroute to Korea; 1942-1943 USS *Maddox* (DD-622); USS *Reno* (CL-96); WWII; USS *Arneb* (KA-56); Korean War 1950-1953; USS *Missouri;* USS *Naifeh* (DE-352); USS *Hancock*.

He was awarded the Asiatic-Pacific w/4 stars, WWII Victory, Korean Good Conduct and American Campaign, EAME w/star.

Moyer was discharged Jan. 15, 1964 as BTCM after 22 years. He has a daughter and two grandsons. He owns his own company, Combustion Service and is semi-retired, enjoying golf, fishing and travel.

ROBERT D. MUNOZ, born Jan. 26, 1926, joined the USN June 6, 1944. He was assigned to USS *Missouri* February 1945 and was on the staff of Adm. Bull Halsey Jr. He served on the signal bridge, conn, 20mm antiaircraft gunner.

While at sea he was a witness of the surrender. His memorable experience includes being struck by friendly fire at Iwo Jima and Okinawa.

He was awarded eight Battle Stars, the American Theatre, Asiatic-Pacific, Philippine Liberation. He served on battleships *South Dakota, Massachusetts, Quincy, Louisville*. He was discharged June 6, 1946.

Muñoz is married to Marjorie and they have three daughters, two sons and 13 grandkids. He is retired and is a traveler and photographer.

WARD R. MUNSON, LT, USNR, volunteered and received a commission of LTjg in the USN, leaving a wife, nine month old daughter and two-year-old son. After indoctrination and communication training, he was assigned to Adm. Nimitz's staff. One year later Capt. Callaghan secured orders for Munson to follow him to the USS *Missouri* where he took command. *See Special Story Section for more on Ward R. Munson.*

JAMES E. NEWTON, joined the USN in 1950 from Punxsutawney, PA, trained at Great Lakes, then the Navy School of Music, Washington. While on the *Mo* they traveled to England, Norway, Cuba and Panama before shipping to Hawaii and on to Korea.

His regular duty was as a musician, but his battle station was 1st loader on the 40mm. One of his memories is playing for the crew during mess when suddenly GQ sounded and they scrambled to their battle stations. They had no time to grab jackets, gloves or hats and as they stood waiting for those MIGS to come in, they watched their bare hands holding the shells turn purple. They hoped for action just so they could move.

He earned the Good Conduct, China Service, National Defense Service, Korean Service, United Nations Service, Navy Occupation Service Medal and the Republic of Korea Presidential Unit Citation.

Newton mustered out in 1954, attended Gettysburg College before going "on the road" with the bands. While playing the Statler Hilton in Hartford, CT, he met Joyce, whom he married in 1958. He has been teaching instrumental music in the regional school system since 1959. They have two daughters, Sandy and Shari, and two granddaughters, Kristen and Danielle.

THOMAS T.X. NGUYEN (TRUONG X. NGUYEN), born April 20, 1969, Chu-Hai (70 miles south of Saigon and Vung-Tau), Vietnam. Entered the USN Sept. 8, 1989. Battleship USS *Missouri* (BB-63) plaque owner. He served aboard USS *Missouri* in 6th, 4th and H Divs. from January 1990 to January 1992.

His education includes attending Navy Corpsman School and AAS-Opticianry. He studies to become an optometrist/medical doctor.

His memorable experiences include: Nickname "Skip", RimPac '90; Helmsman during watch; general quarters in 5"-38 mounts in the Arabian Gulf; Pearl Harbor 50th anniversary and ports visited.

He completed five years of active service at Naval Hospital, Corpus Christi, TX Oct. 8, 1994 as HN. Awarded National Defense Service Medal, Sea Service Deployment Ribbon, First Good Conduct Medal FPE93097, Navy "E" Ribbon, Southwest Asia Service Medal w/Bronze Star, Combat Action Ribbon, Navy Unit Commendation and Kuwait Liberation Medal.

He is an optician, continues to study medicine, and is a member of the church choir.

ROBERT B. NICHOLS, born Jan. 26, 1926 in Balston Spa, NY, went into the USN March 17, 1944. Took basic training at Sampson, NY, then to Newport, RI for training to be a member of the first crew on the USS *Missouri*. He was on the *Missouri* from the commissioning until he was discharged May 14, 1946 at the Fargo Building in Boston, MA. He was a machine gunner on Gun 49, Group 19 in front of Turret 3. He saw and shot at the two suicide planes that hit the ship. They crashed about 70 feet from where he was strapped in his gun. One on April 11, 1945 and the other on April 16, 1945. He has parts of those planes to this day. He also shot at other planes but they dropped into the sea off the ship. Was in the 8th Div., deck section for a short time then got into the gunnery section.

He was discharged as seaman first class, striking for gunners mate third class. He received the American Theatre Medal, Asiatic-Pacific Medal w/3 Stars, Victory Medal and the Navy Occupation Medal w/Asia Clasp.

Nichols will always remember Sept. 2, 1945 Tokyo Bay. When the surrender ceremony was over the sky, was filled with all sorts of planes, you could just about see the sky. It was a sight he will never forget. He is proud to have been there and fought for his country, to him it is the greatest country in the world.

He is married to Nancy Tanguay and they have one son and two daughters, five grandsons and four granddaughters. He retired from Morrison Berkshire Machine Company of North Adams Massachusetts after 18 years as a field service technician on March 1, 1991.

WILLIAM H. NORBERG, born Nov. 28, 1931, Boston, MA, enlisted July 2, 1951 until June 22, 1955, served December 1951 to February 1955 upon the decommissioning of the USS *Missouri* in Bremerton, WA. He was transferred to USS *Hancock* (CVA-19). He received Korean Service and Good Conduct Medals.

His memorable experiences: While in Korea, the spotting helicopter was shot down and the loss of two crew members were stored in the refrigeration spaces

until returning to Japan. Thoughts were of the two families who suffered the losses. He was a cook for 3,000 men.

He is married to Sheila Regina Drummey and they have two daughters and one son. He worked 35 years as a plumber in the Boston area and is currently enjoying retirement.

GARDINER L. NORTHUP, born June 19, 1920 in Providence, RI, joined the USN Aug. 8, 1942, assigned to the USS *Missouri* June 11, 1944 as fire control and achieved the rank of FC3c.

While at sea he participated in Iwo Jima, Okinawa, and Japan home islands.

His memorable experiences: camaraderie among shipmates; lousy food at sea; constant General Quarters.

He was discharged Aug. 18, 1945 and was awarded the American Theatre, Asiatic-Pacific w/3 stars, WWII Victory Medal and Japan Occupation Medals.

Northup retired from Amic Mutual Insurance Company after 40 years in 1980.

He is married to Martha Ohsberg and they have two sons, two daughters and seven grandchildren.

JOHN (JACK) NORTON, born Oct. 18, 1919 in Hoosick Falls, NY, joined the USN and was assigned to the USS *Missouri* during the summer of 1944. Aboard the *Mighty Mo* he served in K Radar and kept his most memorable experiences documented in a diary.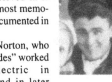

After WWII, Norton, who was a "jack of all trades" worked for General Electric in Schnectady, NY and in later years was self-employed.

Jack Norton passed away Jan. 2, 1965. He left his wife Bernice and their five children: Judith, Sharon, Deborah, Jack and Geoff. Today his family has expanded to 14 grandchildren, nine great-grandchildren and is still growing.

WILLIAM OBITZ, born Sept. 1, 1926 and joined the USN March 17, 1944 and went to boot camp at Sampson, NY. After four weeks he went to Newport, RI for the precommissioning crew of the USS *Missouri*, from there they reported for the commission on June 11, 1944 and served aboard until his discharge on May 12, 1946.

He was a deck hand in the 8th Div. and the 20mm guns crew on the starboard side, turret No. 3 where the suicide plane crashed April 11, 1945. He also was a boat coxswain for the liberty boats, discharged as a S1c, received the Pacific Theatre Ribbon w/3 stars, American Ribbon and Victory Medal.

After his discharge he met Helen and they were married Nov. 7, 1947. Their family includes a son, two daughters and five grandchildren.

FELIX S. OLIVA, born Jan. 14, 1923, Waipahu, Oahu was inducted in USN May 5, 1942.

On November 1944 he boarded the USS *Missouri* in San Francisco, CA and was assigned to ship company under Capt. Callaghan.

Upon arrival in Honolulu, Adm. Halsey came aboard and made the USS *Missouri* the Flagship of the 3rd Fleet. He then became the personal steward to Adm. Halsey and his staff, Adm. Carnie, Commo-

dore Boom, Capt. Cross, Capt. Taylor, Capt. Sloan Commander and Harold Stassen (member of the UN).

Reno, NV presented a Silver Saddle to the USS *Missouri*. Which then became his special seat while watching movies.

After the signing ceremony of the surrender document a Russian General gave him a bottle of cognac to serve to General MacArthur, General Wainwright, Adm. Nimitz and Adm. Halsey.

Admiral Halsey transferred the entire 3rd Fleet on the USS *Dakota* (honoring her heroic battles) to parade the entire fleet to Honolulu and then to San Francisco, CA.

THOMAS J. ONDRAKO SR., born Binghamton, NY on April 2, 1926, enlisted in the USNR Feb. 17, 1944 and went to Sampson, NY for boot training. He attended gunnery school aboard the USS *Wyoming* in Norfolk, VA and from there traveled to Newport, RI and boarded the USS *Chilton* bound for the USS *Missouri*, commissioned in New York on June 11, 1944.

He joined the 4th Div., manning the five inch 38 gun crew when needed, and became a seaman first class. Daily tasks of scraping, priming and painting were done to keep the *Missouri* immaculate.

Their ship engaged in bombarding Kyushu during the Iwo Jima campaign. Easter Sunday found them relentlessly attacking Okinawa. He witnessed the historic signing of the surrender in Tokyo Bay on Sept. 2, 1945. On October 27, Navy Day, New York welcomed the *Missouri* and fleet up the Hudson River, an incredible site. In March 1946 they escorted the remains of the Turkish ambassador through the Dardanells to Turkey, also visiting Greece, Rome and Algiers.

He was honorably discharged on May 13, 1946 as S1c at Lido Beach, NY receiving the American, Asian and Victory Medals w/3 stars.

He became a printer with the ITU for 45 years, married Stephie Jungmann on Jan. 23, 1954. They have four children: Sharon, Tom, Joe and Jim who are all married and have children of their own (eight in all). Stephie passed away on Jan. 26, 1988. He is now retired and enjoys any opportunity when he can visit his children and grandchildren who live in various parts of the country.

IRVING C. OSTERHOUDT, born May 24, 1923 in Hyde Park, NY, married Peg MacLeod in Poughkeepsie, NY, Nov. 14, 1942. He joined the USN Dec. 17, 1943 and was assigned to the USS *Missouri* as WT3c.

He saw action in the Asiatic-Pacific w/3 stars. Though he had many memorable moments aboard ship he has to say the most meaningful was the signing of the Peace Treaty in Tokyo Bay. He was discharged Jan. 5, 1946.

Today he is retired, residing in Florida, working part time, playing golf and traveling. He and his wife, Peg, have a daughter, Linda and two sons, Kenneth and David, 11 grandchildren: Brooke, Shelly and

Charles Levy, Corey, Amy and Jeffrey Osterhoudt, Elaine, Edward, Ethan, Elliott and Evan Osterhoudt and one great-grandchild, Nicole Roberge.

LEO J. O'TOOLE, born Oct. 4, 1926 in Camden, NJ, was inducted into the service July 27, 1944 and assigned to the USS *Missouri* Sept. 4, 1946 as signalman.

His memorable experiences include: while he was on signal watch the *Missouri* was hit by friendly fire from USS *Little Rock* during operations in Argentia Bay with USS *Little Rock* and USS *FDR*.

He was awarded the Asiatic-Pacific, WWII Victory and the American Theatre and discharged Sept. 10, 1947 as SM3c.

O'Toole has a daughter Patricia and four grandchildren: Joe, Veronica, Chris and Peter. He is retired from IBM.

VERNON (VERNIE) OVERTON, born Sept. 21, 1907 in Stoughton, MA, joined the USN, out of Boston on St. Patrick's Day, March 17, 1944. Most of Vernie's USN buddies called him "Pops" on the *Missouri* because he was an old 37 when he was piped aboard ship in 1944. "Pops" proudly served as a seaman first class and was stationed at Gun Turret Number Two on *Big Mo*.

On Jan. 30, 1946, he received his discharge and returned home to his family in Stoughton.

On Sept. 20, 1997, Overton celebrated his 90th birthday with 80 family members and friends on board the battleship USS *Massachusetts*.

Talking sports or politics continues to be Vernie's "tonic." He remains forever young when he explains why the Red Sox lost a ballgame or if a local Boston politician will win an upcoming election.

He is a great-grandfather many times over and happily lives at his home in South Easton, MA with his family.

ROBERT E. (BOB) PARKER, born June 7, 1924 in White Bear Lake, MN, joined the USN Nov. 10, 1942 and was assigned to USS *Missouri* July 10, 1944 as signalman.

His memorable experiences include being on board at the original commissioning and at the recommissioning in San Francisco. He was nicknamed Wahoo by his shipmates. Aboard the USS *Custer* when it was commissioned.

He was discharged Jan. 9, 1946 and was awarded the American Theatre Medal, Asiatic-Pacific Medal w/3 stars and the Victory Medal.

Parker was married to Marqueritte Vadnais in 1948 and they have nine grandchildren, one stepgrandson and one step-great-grandson. He is a retired electrician. Active in White Bear Lions Club, Melvin Jones Recipient VFW #1782. Although he has emphysema he still gardens, makes jewelry, paints, cans hundreds of quarts of vegetables each year. Although his battery has run down he makes the most of every day. They can only hope they'll still be around for the reunion in 2001 in St. Paul.

WILLIAM D. PATTERSON, born Dec. 24, 1928 in Chicago, IL, enlisted in the USMC June 1946 immediately after graduation from high school. After basic training at Parris Island he was assigned to duty on the USS *Missouri*. While in port he was responsible for the captain's car. The *Missouri* took several tours, one of which was to South America. He was aboard in September of 1947 when the President and Mrs. Harry S. Truman and their daughter Margaret came aboard in Rio de Janeiro, on their return to Norfolk, VA. It was on this trip that the Truman's crossed the equator and were initiated from Polywogs to Shellbacks.

A memorable experience during a training exercise on the deck of the *Missouri*, off the coast of Cuba, was an explosion in a 16-inch gun turret, resulting in the loss of several fingers on his right hand. He was sent to Cuba and from there to Jacksonville, FL and back to the States for medical discharge in September of 1948 as private first class. After attending the University of Illinois he made his home in Waukegan.

He ran a vending company in Waukegan, IL for many years before his retirement to Mesa, AZ with his wife Elizabeth. He is a father of four, grandfather of seven and great-grandfather of one.

Four generations of Pattersons' have served in the Marine Corps. His father, Bill, his son and now his grandson.

WALTER L. PAUL, born June 19, 1925 in Summit, NJ, entered USN on Oct. 15, 1943.

His stations include: Newport, RI and Radio School, Boston, MA.

Assigned to USS *Missouri* Naval Yard Brooklyn, NY until discharge Dec. 26, 1945 at Lido Beach, Long Island, NY.

He is a member of the CR Div. as RM3c, receiving coded messages in radio room.

Paul received the American Theatre, Asiatic-Pacific w/3 stars, WWII Victory, Occupational, Service WWII and Good Conduct Medals. Participated in shakedown exercises in Gulf of Paria, Trinidad and British West Indies. Participated in first Tokyo, Iwo

Jima and Okinawa Fast Carrier Task Force strikes. Member of the 3rd Fleet Landing Force which occupied Yokosuka Naval Base, which was his most memorable experience.

He is married to Marion A. Muldowney and they have two daughters, Carolyn and Katherine, and a son Walter Jr. and three grandchildren: Meredith, Matthew and Gwendolyn. Retired from New Jersey American Water Works after 44 years of service. Paul has also been a part time court officer for the past 50 years.

He resides at his winter home in West Palm Beach, FL (four months) and Summit, NJ home (eight months). His hobbies include golfing, swimming and walking.

WALTER D. PHILLIPS, born April 13, 1921 near the small town of Macclesfield, NC, joined the USMC Nov. 12, 1942.

He was awarded the American Campaign Medal, Asiatic-Pacific Campaign w/5 stars, Philippine Liberation w/2 stars, WWII Victory.

His memorable experience was the privilege to serve under Adm. Halsey, a truly compassionate commander to all he commanded. Also having his brother Sgt. Willie T. Phillips being one of the 13 Marines to service along with Walte Phillips on Adm. Halsey's Flag. And last being a participant in the official ceremony of the Peace Treaty on the *Missouri* Sept. 2, 1945.

Phillips married his high school sweetheart in 1943, while stationed at the Naval Mine Department at Yorktown, VA. His wife worked at the Naval Payroll Office until he left for overseas duty in the South Pacific on Nov. 15, 1943.

After serving in the Marines, returned to the family farm November 1945.

He had two sons, four granddaughters and four great-grandchildren. Retired from farming in 1984 and is presently employed at the Maccripine Country Club and Golf Course.

WILLIAM R. PILLSBURY, born July 5, 1924 in East Braintree, MA, joined the USN March 17, 1944 and was assigned to the USS *Missouri* June 11, 1944. His battle station was on top of turret #3 on the 40mm guns, deck crew on the fantail of the ship.

While at sea they were in Task Group 58.2. They bombarded Iwo Jima, Okinawa and then mainland of Japan.

His memorable experiences include getting hit by a Jap kamikaze plane which had already dropped his bomb. The surrender ceremony aboard their ship on Sept. 2, 1945.

He was discharged May 13, 1946 as S2c in 11th Div. and received awards for the American Campaign, WWII and Asiatic-Pacific Campaign.

Pillsbury is married and lives in Whitman, MA. He is retired from the Braintree School Dept. and worked in the auto parts business also.

LOUIS S. PISANI, born Nov. 29, 1924, entered the USN March 1943. His stations include: Great

Lakes, May 1943 USS *Lynx* South Pacific; June 4, 1944 USS *Missouri*. Discharged Jan. 30, 1946.

Served on USS *Lynx* 5"/51 loader supply and amphibious support, USS *Missouri* FA Div., target designator air defense. Good location to observe the kamikaze attacks, the planes that hit them, USS *Franklin* and others. Job Mark/14 maintenance. Assigned A Div., a motor mac.

Served USNR to 1943. Enlisted Army Reserve, activated 1961-1962 Berlin Crisis, 1st Sgt. Arty. Btry. Promoted to command sergeant major, Arty. Bn. Finished his career as director NCO Academy.

He was awarded the Navy American Defense, Asiatic-Pacific w/3 stars, Victory, Occupation Medals Army, Meritorious Service, Army Commendation, National Defense, Armed Forces Reserve, Army Reserve Achievement and Army Service Medals.

He married Frances Lasher and they have two sons, two daughters and three grandchildren. Retired from the GTE Corp and the US Army.

WILLIAM A. PITCHER,

enlisted in the USN in the V-7 Program in August 1940. He was released after the midshipman cruise and reentered the program in March 1942. Commissioned in February 1943 and served on the USS *Elizabeth C. Stanton* (APA-69) and the USS *Bunker Hill* (CV-17) during WWII.

After duty in the NROTC unit of the University of Pennsylvania, he was accepted into the Regular Navy and ordered to the USS *Missouri* (BB-63) in October 1946. Due to previous experience he was assigned as R Div. officer and fire marshal, later moving to the 4th Div. officer and 5" gun battery officer. Ordered from the *Mo* in February 1948 to Line School, he went on to serve on the USS *Ozbourn* (DD-846), USS *Herbert J. Thomas* (DD-833) and USS *Vesuvius* (AE-15) during the Korean and Vietnam conflicts and beyond. He served as CO on the later two ships.

He also served two tours on Joint Staffs, Cin Carib, and US Forces Korea and Deputy Supt., US Naval Post Graduate School.

Pitcher retired as captain on May 1, 1973, recipient of 20 medals and decorations highest being Bronze Star, w/Combat V.

Widower in 1992 after nearly 46 years of marriage, he has two sons, both whom served in the USN, their wives and four grandchildren.

He lives in Danville, CA and enjoys playing golf at nearby Diablo Country Club.

DOUGLAS C. PLATE,

arrived pre-commissioning detail of USS *Missouri* in April 1944 via 5" and 16" gunnery school. Served as 5" division officer and catapult officer during shakedown and early part of Pacific duty. Transferred to main battery duties in February 1945. MK was in 37 Director Cond III, and MB Plotting Room Officer Cond I. He was junior officer of the watch for surrender ceremony and senior watch officer for return trip to New York. Detached November 1945.

Highlights: Transit of Panama Canal, Ulithi Atoll, director officer during shoot down of three Japanese aircraft; bombardment of mainland Japan, July 1945; OOD for morning watch which received 37 ships alongside, Aug. 12, 1945; transfer of 700 men to *Iowa* via high line August 28, "Light ship"; JOOW for surrender, Sept. 2, 1945; OOD for Navy Day ceremonies in Hudson River with President Truman, Governor Dewey of New York, Gov. Edge of New Jersey, Secretary Forrestal, Adm. Leahy, etc. First peacetime gun salute.

Plate married Peggy Deane in 1945 and they have two children and four grandchildren. He retired in 1975. His best memory was a lot of wonderful guys as shipmates.

CLYDE E. POPE,

born July 14, 1925 in Brookford, NC, entered the USN in June 1943. His stations include: Perryville, MD; Newport, RI. Assigned to USS *Missouri* after her commissioning in March 1944, Brooklyn Navy Yard. Member of the 8th Div., 20mm guns, was coxswain.

His memorable experiences: kamikaze planes attack at Okinawa; the typhoon somewhere in the Pacific, 1945; witnessing the signing of the surrender of Japan in Tokyo Bay, Sept. 2, 1945; playing on the ship's baseball team after the war's end; and Navy Day in New York, October 1945 when President Truman reviewed the fleet. March 1946 was date of discharge.

He earned the Asiatic-Pacific Area w/3 stars, American Theatre, WWII Victory and Good Conduct Medals.

Pope received an AB degree from Lenoir Rhyne College, Hickory, NC in 1950 and M.Ed. from UNC-Chapel Hill in 1955. He married Ruth Soeldner in 1948 and has five children and five grandchildren. Retired from Charlotte-Mecklenburg Board of Education after 33 years as teacher, principal, and administrator. He is an elder in the Presbyterian church in which he is very active. He gardens, travels, and is enjoying life.

ALOYSIUS ROBERT PROSAK,

born Aug. 20, 1926 in Lorain, OH, joined the USN Aug. 25, 1944 and was assigned to the USS *Missouri* Dec. 1, 1944 serving in refrigeration A-Div.

While at sea he participated in Iwo Jima Feb. 19-23, 1945; strikes of Tokyo Area, Feb. 25, 1945; strikes of Okinawa Shima, March 1, 1945.

His memorable experiences include the Japan surrender Sept. 2, 1945; Ulithi, Anchorage and the Panama Canal.

Admiral (Bull) Halsey Third Fleet hoisted his flag aboard USS *Missouri* on May 18, 1945 in Guam.

He liked to walk on deck at chow time, talking with crew members. Ltjg. Swack A Div. boss was a wonderful person. March 1946 they sailed to Istanbul, Turkey with the Ambassadors body. They then traveled to Athens, Greece, Naples, Italy. Visited Rome as audience with Pope Pius on Good Friday. It has been a long time since a Pope had visited on Good Friday. They visited Algiers and Tangiers. He enjoyed his time on the USS *Missouri*.

Prosak was discharged June 5, 1946 as F1c. He received the Victory Medal, Asiatic-Pacific Area Campaign w/3 stars and the American Area.

He is married to Dorothy and they have three sons and five grandchildren. He retired and enjoys golf, fishing and helping the high school booster club.

BOB L. RAFFERTY,

born July 4, 1925 in Sioux City, IA, joined the USN Sept. 9, 1942 and was assigned to the USS *Missouri* February 1944. Also assigned to No. 1 Control Benchboard, No. 1 Engine Room, as electrician's mate. February 19, 1944, the ship shot down its first plane. He was discharged Oct. 26, 1945 as EM2c.

The highlight of his career was being present for the signing ceremonies of the surrendering of the Japanese, and seeing Gen. Wainright able to attend the same.

On May 10, 1986 he was in San Francisco for the second commissioning of the USS *Missouri*. He was very proud to have been a crew member of this fine ship. He passed away in January 1997.

OTTO F. RE,

born July 2, 1925 in Mount Vernon, NY, joined the USMC in October 1943. After boot camp, Parris Island, he was assigned to the Marine contingent aboard the USS *Missouri*.

His memorable experience: one day while walking aft of the starboard side of the ship a voice called out "Marine", snapping to attention with a brisk salute, he answered "Yes sir", it was the voice of Adm. Bull Halsey who asked if he was on duty, "No sir", Re replied, after which the Admiral asked "Would you like a fast game of handball?" Re answered, "Yes Sir." Who would ever think he played a game of handball on the very spot where the formal surrender was signed, on the starboard side of the 16' gun turret No. 2.

Bad experience: while manning the gun station, 20mm gun on the starboard side, opposite 16' turret #1, they were attacked and hit by a kamikaze whose wing landed about 10 feet past the gun station on the deck. When it was said that an experience such as this causes your life to flash by in the blink of an eye, truer words were never spoken.

He is married to Rose Di Gregorio and they have two children and is employed by Key Cadillac in White Plains, NY.

MARTIN J. REILLY,

born Sept. 23, 1909 in Brookline, MA, drafted in August 1943 and assigned to the USS *Missouri* Aug. 13, 1943 where he held a rank of machinist mate third class. He was on the *Missouri* until his honorable discharge, Oct. 26, 1945.

His greatest thrills were at battle in the Pacific and to be present on the historic day the Peace Treaty was signed, Sept. 2, 1945 in Tokyo Bay. He has left those mementos and memories with his four children: Patricia, Edward, Ellen and Martin Jr.

He saw plenty of action in the pacific in WWII

and was often called "Pops" by the younger men, because he was drafted much older.

After his service days, he became very active in the VFW Post Stephen Rutledge in Brooklin, MA for more than 35 years, servicing the local community, until his death on April 15, 1982.

His wife Mildred, died Dec. 9, 1991. They are survived by their four children and six grandchildren.

WILLIAM FRANCIS REILLY, born Jan. 13, 1919 in Jamaica, NY, joined the USN Jan. 3, 1942 as yeoman third class, after schooling at Princeton University. His rating was changed to gunner's mate third class. He was stationed in England and Germany and was active in the invasion of France.

He received five medals and a Bronze Star and was discharged Oct. 4, 1945 as GM1c.

Reilly was recalled for the Korean conflict Sept. 4, 1951, stationed aboard the *Missouri*. He was very proud to have served on the *Big Mo.* He was discharged Nov. 16, 1952.

He and his wife Constance have two daughters, Elizabeth (Betsy) and Madelyn, three granddaughters: Jennifer Reddin Eliou, Erin and Heather Cummings, two grandsons, William and Sean Reddin. Reilly died June 13, 1980.

FRANK H. REIMERS, born May 27, 1918, enlisted in the USN in July 1942. After a tour of duty in the Atlantic aboard a light cruiser and upon completion of Advanced Fire Control School, Washington, DC, he received notice for active duty aboard USS *Missouri* (BB-63) in April 1944. As firecontrolman second class FA Div., his duties were to bore sight and align five inch gun mounts to controlling director.

One memorable experience was the birth of the *Missouri,* commissioned June 11, 1944. Another incident was during an air attack when a Japanese kamikaze crashed into the starboard side of the ship with a portion of its wing coming to rest just below Sky 3, Frank's battle station. He retrieved a small portion of that wing which he still has today. Never to be forgotten was the signing of the surrender by the high Japanese officials to the Allied Powers on Sept. 2, 1945 which he was very proud to witness.

Today, he is retired from Chrysler Corp. and is living in Scottsdale, AZ with Carol, his wife of 53 years.

CARL REISMAN, born June 11, 1920 in Atlanta, GA, entered active duty July 7, 1941 as an ensign after completion of NROTC at Georgia Tech. Served on board the old heavy cruiser USS *Quincy* engaged in convoy duty in the Atlantic. Commissioned and plank owner of the following three ships on which he served consecutively in the antiaircraft division: USS *Columbia* (light cruiser), USS *Iowa* (BB-61), and the USS *Missouri* (BB-63). He was on board the USS *Iowa* when President Roosevelt and his staff went to meet with Churchill and Stalin at Teheran and Roosevelt's return to the US. He also was ordered by Adm. Halsey to pick up the Japanese surrender party and bring them

to the USS *Missouri* on Sept. 2, 1945. He was detached from the USS *Missouri* as a lieutenant commander on Oct. 20, 1945.

He was awarded the American Defense Service Ribbon, with "A" for duty on USS *Quincy,* American Area Ribbon, EAME Ribbon, Asiatic-Pacific Area Ribbon w/4 stars, Navy Occupation Service Medal (Asia) and Victory Ribbon.

He was married to his late wife Betty Scholer for more than 43 years until her death in July 1987. He has a daughter, a son and three grandchildren.

Rusman married Claire Wertheim in May 1990, and has two stepsons. He retired in 1987 as CEO from Tri-anim Health Services, Inc., Sylmar, CA. He and his wife reside in Encino, CA.

EDWARD FRANCIS RICHARDS, born Feb. 26, 1926 in Staten Island, NY, joined the USN March 24, 1943 and assigned to *Missouri* June 1944 (plankowner), 20mm sight setter, achieved the rank of seaman first class.

While at sea: three engagements on USS *Chester* (CA-27); three engagements on USS *Missouri* (BB-63).

His memorable experiences: On Sept. 2, 1945, he was assigned to the *Missouri's* motor launch. They were instructed to man their boat and lay off the side to await orders to pick up the Japanese delegation and bring them back to the ship. At the last minute it was decided to use the motor launch assigned to the USS *Iowa.* They had to lay off the starboard side of the ship and watched the signing of the surrender from there. It was a great view.

He was discharged Jan. 30, 1946 and awarded the Asiatic and Pacific Theatre w/6 stars, American Theatre and the Victory Ribbon.

Richards is retired from Kraft Foods after 32 years in sales and now a recording secretary of the USS Missouri Association.

He has been married to Eileen for 48 years and they have one daughter, Denice, two sons, Mark and Neal and one grandson, James Patrick.

JERALD EUGENE RICHARDSON, born Nov. 20, 1921 in Britt, IA, joined the USN Nov. 5, 1940 and was assigned to the USS *Missouri* February 1944, as deck hand, cox. He was aboard the USS *Arkansas* for four years. He received the American Defense Service w/A, American Area, EAME w/star and the Good Conduct Award and was discharged Dec. 21, 1946.

Richardson retired after 40 years with John Deere Tractor.

FRED PATRICK RILEY, born April 24, 1925 in Omak, WA, enlisted in the USN Aug. 10, 1942, attended basic training in San Diego, CA; Machinist Mate School, ND and transferred to USS *Harris* (APA-2). Assigned to engine room and promoted from F1c to MM2c.

Participated in troop landings Aleutian Islands and Tarawa Island. Reassigned to USS *Missouri* under construction, Brooklyn Navy Yard, January 1944.

Assigned to *Missouri* main control engine room until separated as MM1c May 16, 1946.

Recalled to active duty from reserves as MML1 September, 1950-December 1951. Returned "mothballed" ships to service in San Francisco and Long Beach.

He was awarded the Navy Unit Commendation Ribbon, Good Conduct Medal, American Campaign Medal, Asiatic-Pacific Medal w/Silver Star, WWII Victory Medal, Navy Occupation Medal w/Asia Clasp and the National Defense Service Medal.

Riley was married to Eileen Fitzpatrick for 50 years, Dec. 31, 1996. They have one daughter, a biologist and two sons who are police officers. He retired from electric utility after 37 years of service. Attributes satisfying engineering career to USN training.

H. KIRBY RINER, born Dec. 1, 1913 in Funkstown, MD, moved to Berkeley County, WV shortly thereafter and has been a life long resident. High school and business college graduate. Enlisted in the USN February 1944 and served aboard the USS *Missouri* until discharged as AMM3c on Dec. 6, 1945 at the war's end, receiving various medals.

He was married to Louise Hite Riner for 48 years and she is now deceased. He has been married to Mary Jean Honn for 13 years. He has three daughters: Sharon Riner Lewis, Rebecca Riner Masters and Christine Riner Holben and three stepchildren and a total of eight grandchildren, eight stepgrandchildren and five great-grandchildren. Riner retired as a foreman from Martin Marietta in 1976 after 41 years of service.

THOMAS E. RISCH, born May 27, 1930 in Anthony, KS, enlisted in the USN with 12 friends after high school in 1948. He took boot camp in San Diego, then met the *Missouri* at Norfolk, VA.

Enjoyed seeing Oslo, Norway and Cherbourg, France on midshipman cruise. He was on the *Missouri* aground in Hampton Roads, VA. September 1950, crossed the International Dateline on the trip to Korea. Christmas was spent at Hungnam and was brightened by Bob Hope and troupe.

He served under five captains while aboard, as a storekeeper in clothing and small stores. He served under commander J.N. Witherall.

Risch married Charlotte Bishop in 1949 and had one son and one grandson. He was an auditor and station manager for Mobil Oil, died Dec. 18, 1995.

He is always proud of the *Mo.* They attended the decommissioning ceremony March 1992 in Long

Beach and 50th anniversary at Bremerton, WA September 1995.

MARLIN E. RITTER, born Sept. 11, 1922 in New Berlin, PA, entered USN March 26, 1943, NTS Sampson, NY, NTS, MM. Wentworth Institute Boston, MA.

Assigned to the USS *Turner* (DD-648) of the Atlantic Fleet on convoy duty. The ship was lost Jan. 3, 1944 and he was lucky to be a survivor.

Ritter was assigned to the USS *Missouri*, Brooklyn Navy Yard as nucleus crew in B Div.

He was present for precommissioning of the ship, June 11, 1944 and spent the entire tour of duty in #4 fireroom as a WT2c. He was also present for the signing of the treaty.

Ritter was married to Miriam Miller for 53 1/2 years. She passed away Dec. 11, 1995. They had two children, George and Karen. He retired from the US Department of Justice Prison System in 1985.

HENRY C. RIVERS, born Sept. 7, 1923 in Worcester, MA, received his elementary education in Hubbardston, MA. Graduated high school 1941, Athol, MA. Enlisted in the USN July 1, 1941 and completed recruit training at Newport NTS, Newport, RI. He enlisted under the name Henry C. Rivinoja. His name was legally changed to Henry C. Rivers on Jan. 21, 1946.

His ships and stations: September 1941-April 1942; advanced base aviation training units-NAS Norfolk, VA and NAS Quonset Point, RI. He participated in fabricating and erecting the first quonset huts which became common to all service branches. April-June 1942: Sub Chaser Training Center, Miami, FL; June-July 1942: pre-commissioning and fitting out crew USS *SC-656* in Rockland, ME; Commissioned USS *SC-656* at Boston, MA Aug. 11, 1942; Operated out of Convoy Control Center, Key West, FL during 1943 and 1943 doing convoy duty and antisubmarine patrols in Caribbean and South Atlantic Ocean; Attained the rating of coxswain; September 1943, Lighter-than-Air School, Lakehurst, NJ; December 1943, Assigned to Blimp Headquarters Sqdn. 4 at Fortaleza, Brazil; Attained rating of BM2; November 1944, transferred to receiving station, Riverside, CA (San Francisco); December 1944, assigned to USS *Missouri* (BB-63); 10th Div. 1944-1945, Mk51 director operator (40mm); January-April 1946, 7th Div. April 1946 until discharge Oct. 30, 1946: XO Div. (MAA Force).

While at sea he participated in action at Iwo Jima, Okinawa, and Japanese Home Islands. His awards include the American Defense, American Campaign, Good Conduct, Asiatic-Pacific Campaign w/3 stars, Philippine Defense/Liberation/Independence (Philippine Issue), Asia Occupation and Victory Medal. He was discharged as BM2c Oct. 30, 1946.

His college degrees include an AA, Worcester Junior College; BSBA, Clark University, Worcester, MA; CLU, American College, Bryn Mawr, PA.

Rivers retired as an insurance/investment broker. He married Vera V. Rivers Nov. 23, 1945. His memberships include the USS Missouri Association, Inc., American Battleship Association, Patrol Craft Sailors Association, American Legion Post 306, Veterans of Foreign Wars Post 15022, United States Naval Institute, US Navy Memorial (plankowner), American Society of Chartered Life Underwriters and Chartered Financial Consultants, Golden Key Society, United States Historical Society, Finland Society and Finnish Cultural Society.

WILLIAM PAUL ROBBINS (BOLESLAUS PAUL RYBINSKI), born Jan. 26, 1928 in Buffalo, NY, enlisted on his 17th birthday, Sampson NTC, New York, USS *Missouri* (BB-63), as seaman second class, R Repair Div. while assigned to three months mess cooking. It was his mess that was used by General MacArthur to have the Japanese delegation sign the instrument of formal surrender ending the second World War on Sept. 2, 1945 in Tokyo Bay, Japan. The table was brought back to him in his mess hall and was mixed with the other tables, it was never tagged or saved for posterity.

Another memorable experience was a conversation with Adm. William F. Halsey. The Admiral commented on his age (17) and his duties in "R" Div. while he was doing repairs in his quarters.

Robbins was awarded the American Theatre, Asiatic-Pacific w/2 stars, Victory, Philippine Liberation and Philippine Independence.

During the Korean War he served as sergeant first class with the 27th Inf. Div. New York National Guard.

He and his wife Jean have two daughters, Candace and Mary Louise and two grandsons, Thomas and Andrew. Jean's father, John Lukowski petty officer first class served on the presidential yacht, USS *Mayflower* under three presidents.

He attended the University of Buffalo in engineering. He held position of division engineer, and retired after 44 years of service with the New York Central, Penn Central and Conrail Railroads in May 1990.

JOHN J. ROHM, reported aboard the USS *Missouri* Dec. 25, 1952 for duty with the Flag Allowance of Vice Adm. J.J. Clark, Commander 7th Fleet. The Flag Allowance of Com7thFlt was transferred to the USS *New Jersey* on April 6, 1953.

ROBERT LEE ROUTH, born Oct. 25, 1927 in Cincinnati, IN, joined the USN July 25, 1945 and was assigned to the USS *Missouri* March 20, 1946, as fireman in the boiler room. His most memorable experience was on the Turkey cruise, visiting six countries.

Routh was discharged Oct. 23, 1947 and received the Victory Medal and American Area Medal.

He is retired after working 43 years in a GM factory. He exercises daily by walking, and does volunteer work. He and his wife, Darlene have two sons,

Mike (deceased) and Jeff, one daughter, Shari, three grandsons: John, Charlie and Shawn, two granddaughters, Kelly and Katie and two great-granddaughters, Ashley and Nikki, and one great-grandson, Christopher.

RICHARD (DICK) ROUTSON, born Jan. 14, 1926 in Flint, MI, joined the USN Dec. 28, 1943 and attended Great Lakes, IL boot camp. Attended school for basic engineering.

He was assigned to the USS *Missouri* while still in drydock April 1944. They stayed in the Brooklyn Navy barracks until commissioning. He was assigned to Engine Room No. 3 and became F1c. He worked the throttle for the engine.

He was at Tokyo Bay at the signing of the surrender, then made the trip to Istanbul, Turkey. Served aboard the *Missouri* 1944-1945 and was discharged May 6, 1946 and received three Pacific Stars.

He owned a resort in Hawaii and retired in 1986. He is now living in Brookings, OR with his wife Janet. They have three daughters, one son and one great-grandchild.

STANLEY RUBENSTEIN, born Aug. 24, 1927 in Brooklyn, NY, enlisted in the USN several weeks after graduating high school as WWII was ending. Attended boot camp at Bainbridge, MD, a short stay at Great Lakes and balance of enlistment as disburser in New York. Discharged September 1946 and remained in Reserves for four years.

Upon completing the City College of New York June 1950, enlisted again as the Korean War began. Boarded the USS *Missouri* three days after enlistment, remaining on the *Mo* from August 1950 until discharge October 1951, as storekeeper third class. After serving in Korea for six months, spent several months on midshipmen cruises to Norway and France.

Two memorable experiences: It was the winter of 1950 and the weather was the coldest ever experienced, since they were not far from Vladivostok. Their bodies were layered with clothes, with only their eyes and nose uncovered. Since this was a war zone (even though it was a police action), their watches were four on and eight off. His duty was a 40mm gun and this night, two of them were on watch for mines, one with binoculars and the other with earphones, in contact with the officer in charge, located either on the 05 or 06 level. Because of the ultra cold weather, both of them located to the warmth of the smoke stacks, for momentary relief. During their absence, the officer in charge tried to contact them. Upon returning to the mount, they were commanded to join the officer.

Two phobias have been part of his psyche since birth: snakes and height. At this moment, he would have welcomed any encounter with the most deadly snakes rather than climb the ladder. Barely entering his mind was leaving their post in a war zone, punishable with desertion. (They do not give Medals of Honor for such acts). After spending an hour up there, and with the fear of height and thought of a court-martial, the cold weather became a minor annoyance. Fortunately, the officer took no action, as they both returned happily to their frigid post.

Serving as the flag ship of the 7th Fleet, they had occasion to host several dignitaries. After racing from the east coast of Korea to the harbor of Inchon, Gen. MacArthur received the grand tour of the *Missouri*. All the Marines that served aboard the ship were reviewed by the general. Each Marine held his rifle perpendicular to the deck and was quite an impressive show. Then, without any warning, MacArthur lunged at one of the Marines, grabbed his rifle, and dragged him several feet.

Having been aboard ship less than two months, his mind set was still locked into civilian behavior, in which such action could be interpreted as deranged behavior. He was convinced that their WWII hero of the Pacific was in desperate need of psychiatric help and perhaps two months of R and R. It was then later explained to this civilian-sailor that Marines never let go of their rifles. To do so could be punishable with a court-martial. You got to know the rules of the game.

After discharge he was employed several years in retailing, then 35 years in education, with a masters degree plus 60 credits. He is married to Barbara Bryant and they have five children: Courtney, Jamie, Kimberly, Jordan and Sharon, plus four grandchildren.

LEROY (HUCK) RUBERY, born Sept. 28, 1923 in Nescopeck, PA, joined the USN Jan. 11, 1943 and was assigned to the USS *Missouri* June 11, 1944. On board he served as quartermaster third class. The action he saw at sea was at Okinawa, Iwo Jima and the bombardment of Maman Iron Works in N. Hokkaido, Japan.

His most memorable experience was working with Adm. Halsey and Lt. Cmdr. Kato, a Japanese pilot who took *Big Mo* into Tokyo Bay.

Rubery was discharged as QMC3c Jan. 17, 1946 and received three Pacific Stars.

He is a retired woodworker, carving ducks, owls, etc. He is also a musician. He and his wife, Iona have two sons, Ronald and Richard; one daughter, Robin; a grandson, Brian; granddaughter, Shannon Price; and a great-granddaughter, Carlyn Wolfe.

LOUIS R. SANDMANN, born Nov. 22, 1931 in Cincinnati, OH, was inducted into the USN May 4, 1952 and assigned to the USS *Missouri* July 1952, 5th Div. Right Projectile man #7. He was discharged March 4, 1954.

He is married to Patricia Powers and they have seven daughters: Cindy, Donna, Cathy, Patty, Nancy, Lori, Becky and nine grandchildren. He is now working at *Cinti Enquirer* a (Gannett newspaper) in the mailroom.

JOSEPH A. SANSO, born March 27, 1926 in New Haven, CT, joined the USN March 14, 1944 and was assigned to the USS *Missouri* just before it was commissioned, 1st loader, on 40mm, with the 10th Div., stationed behind #1 turret.

His memorable experiences include the day they took a kamikaze Jap plane (suicide plane). Commissioning the ship so big he didn't think it would float.

Sanso received the Victory Medal, American Theatre Medal, Asiatic-Pacific Medal w/three stars and was discharged May 13, 1946 as S1c.

He is now widowed and has two fine children, a boy and girl. Retired from the power company with 40 years service. He is now puttering and doing some bike riding.

J. RICHARD SANTO, born Nov. 18, 1932 in Everett, MA, enlisted in the USN May 1950. Attended recruit training in Great Lakes, IL.

While in recruit training, the Korean conflict broke out and immediately upon graduation was assigned to the USS *Missouri,* upon which he served an entire tour of duty.

Served two tours of duty in Korea. The first tour with his friend Sal and the second tour with his younger brother Paul.

Participated in the Inchon invasion and the Hungnam evacuation at Christmas, 1950. He was a rangefinder operator for the main battery. Was involved in shelling operations from Pusan to Wonsan to Chonjin.

Traveled to many countries, the last being Rio de Janeiro, Brazil on a midshipmen cruise. The highlight of the cruise was becoming a shellback and participating in the initiation of the entire USN football team on the return trip.

He was discharged from the USN in September 1953 and attended college in Boston, MA. He is a member of the USS *Missouri* and American Battleship Association.

His awards include the Good Conduct, American Defense, Korean Campaign w/5 Battle Stars, United Nations, Navy Occupation (Japan) and China Service.

On June 19, 1989 he and his wife Marlene and he made a one day sea cruise on the USS *Missouri* for the "sailor for a day" cruise out of Long Beach, CA.

He has been retired since December 1994 after serving as the manager of the real estate department for the city of Anaheim, CA, a position he held for 15 years. He also has been and is still involved in many civic functions, including running for public office.

He and his lovely wife Marlene of 40 years now make their home in the beautiful desert community of La Quinta, CA. They travel extensively in between visits to their children living in Las Vegas and Telluride, CO.

W. ERNEST SAUNDERS, born Dec. 25, 1923 in Sylvester, GA, entered US Naval Academy in 1943 and graduated in June 1945 with the class of 1946.

His most memorable jobs aboard were second division officer and officer of the deck duties.

His memorable events include the Japanese surrender ceremonies Sept. 2, 1945 in Tokyo Bay. The trip home via the Panama Canal.

Various US ports visits, and their cruises to the Mediterranean, South America (with President Truman embarked), and elsewhere. Detached *Missouri* October 1947 with orders to flight training. Had a varied career in naval aviation that included flying Corsairs from carriers, and later assignments in ASW and the amphibious forces. Last USN job was as XO NAS Key West. Retired 1971 in grade of commander.

He is now and has been a fishing guide in Key West since retirement.

Saunders married Dorothy in 1949 and they have six children and seven grandchildren.

ERMINIO SAVELLONI, born Aug. 22, 1925 in Ferentino, Italy, joined the USN Oct. 8, 1943 and after

boot camp at Sampson, NY was assigned to the USS *New York* and after he was picked to start the USS *Missouri* crew at Newport, RI. He served aboard the USS *Missouri* as WT3c the action he saw was at Iwo Jima, Okinawa and the bombardment of Maman Iron Work in N. Hokkaido Japan.

After the armistice, they returned the ashes of the Turkish Ambassador (to the US) to Turkey.

His most memorable moments were on the way back. They stopped at Naples, Italy and he was able to go back home at Ferentino to see his grandparents. He was discharged May 12, 1946.

Today he is retired and with his wife Anna live in Swarthmore, PA. They had three sons: Angelo, Erminio Jr. and John and one grandson and five granddaughters.

JOSEPH C. SCHELL, born Sept. 4, 1920 on the family farm in Crosby, ND, enlisted in the USN at Los Angeles, CA Aug. 14, 1942. He attended boot camp and Hospital Corps School at San Diego.

His stations and ships: San Diego Naval Hospital; North Island; NAS San Diego; Aux NAS, Holtville, CA; Base Hospital #8 Pearl Harbor, HI; Medical Supply Depot, Pearl Harbor; USS *Hamblen* (APA-114); USS *Missouri* (BB-63) May 11, 1946-May 11, 1947; Navy Amphibious Base, Little Creek, VA; USS *LeJeune* (AP-74); USS *Brannon* (DE-446); USS *LST-1146*; USN Recruiting Sta. Seattle, WA.

He was discharged Nov. 20, 1951 as PhM2c and was awarded the Asiatic-Pacific, WWII Victory and Good Conduct.

Following USN service, he worked at hospitals, department stores, and hotels, in Washington, Oregon and Florida and assisted on the farm. A single man, Joe is now retired, residing at Williston, ND, summers, and winters in Arizona.

RAY EDWARD SCHOLL, born April 8, 1917 in Zionsville, PA, joined the USN Sept. 25, 1943 as an ensign. He was assigned to the USS *Missouri* in 1944 as a line officer and joined the crew in Newport, RI. Left the *Missouri* September 1947.

He was discharged from the USN June 30, 1964, worked for Bethlehem Steel Sparrow Point from July 1, 1964 to April 1982.

He and his wife, Doris, had two sons, Ray Grant and Garry Lee and three granddaughters, Sarah, Emma and Anna and died Aug. 29, 1992.

BENJAMIN D. SCHULMAN, born Decatur, AL, 1917, graduated Vanderbilt University, received a BE degree in 1938. In 1940 he volunteered in the USNR. He was commissioned as ensign in 1941 (90 day wonder). Served aboard heavy cruisers, USS *Tuscaloosa* and *Wichita*. He served on USS *Massachusetts* (BB59/plankowner). Served as gunnery officer 1942-1944. Promoted to two striper.

Ordered to *Missouri* (BB-63) January 1944. Reported to Newport, RI. Ship commissioned June 11, 1944. Assigned duties as senior underway watch officer and turret two officer.

Schulman married Ruby Gryzmish in Boston six days later. Sailed into Yokosuka Naval Base (Tokyo) August 1945.

August 18, 1945 he was ordered to new heavy cruiser *Springfield*. Promoted to lieutenant commander USNR as chief gunnery officer.

He lived in Boston in 1977 and moved to La Costa, Carlsbad, CA. He was married to Ruby for 52 years and they had five children. She died in 1996. He is Presently active in video production in Hollywood, CA and owns large mobile television units.

His most memorable events: Sinking the French battleship Sean Bart at Casablanca in November 1942. Kamikaze hitting the aircraft carrier *Franklin,* 1000 dead in minutes.

Best remembered, answering Ward Room Telephone August 1945, and receiving message, Japan officers to surrender if Emperor remains. The war was over. Homeward bound.

ROBERT L. SCHWENK, born 1926 in Glendale, CA, was inducted into the USN in 1944 and assigned to the USS *Missouri* upon commissioning in 1944, B Div.

His memorable experiences include winning the heavy weight boxing championship of the USS *Missouri*. He was discharged August 1947 as MM2c.

He is married and has three daughters, two grandchildren and two great-grandchildren. Schwenk is now retired and has a state record and world record for a white fish caught in 1988 in California.

LLOYD W. SELMAN, born Feb. 18, 1923 in Richardson, TX, enlisted in USMC Nov. 4, 1942 at Dallas, TX, was stationed at Camp Pendleton, Oceanside, CA until being assigned to Asiatic-Pacific Area April 17, 1943. He went to New Caledonia where he was assigned to Adm. Halsey as an orderly. Traveled with Adm. Halsey the remainder of the war.

Left New Caledonia on USS *New Jersey* Aug. 23, 1944-Jan. 27, 1945. After two more ships came aboard USS *Missouri* May 18, 1945 at Guam.

After peace treaty signing ceremony he boarded USS *South Dakota* for trip to states. Arrived Oct. 20, 1945 and was sent to Camp Pendleton, Oceanside, CA. He was discharged as corporal on Nov. 14, 1945. Returned to Dallas, TX.

Many things happened, but the most memorable was witnessing the peace treaty signing ceremony. He was assigned to guard a Japanese emissary. The most tense and scary times was watching Japanese suicide planes attack their ship and others in the fleet. Being

in a typhoon at sea was also an experience not to be forgotten.

His decorations include the Asiatic-Pacific Ribbon w/5 Stars, Philippine Liberation Ribbon w/2 stars, Rifle Sharpshooter, American Theatre Ribbon and Victory Ribbon.

He married Doris L. Covey June 12, 1948 and they have a daughter, Deborah K. Salch; son, Dane D. Selman; a grandson, Brandon L. Johnson; granddaughter, Chelsea L. Nored and two great-grandsons, Blake M. Nored and Hunter L. Nored; one son-in-law, Richard Salch; two stepgrandsons, Christopher and Jeremy Salch; one grand son-in-law, Thomas Nored Jr.

He retired from Koch Refining Company, Corpus Christi, TX December 1989 and now lives on Lake Granbury, Granbury, TX.

EUGENE M. SHIMP, born April 10, 1925 in Akron, OH, joined the USN April 17, 1942. He boarded the USS *Missouri* with the pre-commissioning detail June 11, 1944, serving until Dec. 31, 1945. He worked in the Soda Fountain SS Div.

While on R&R in the Carolinas they were given three cans of beer, which he sold for $1 a can. With that $3, played craps with the Sea Bees running $3 up to $715. A buddy suggested he send some of that home. Good thing! After returning home it was enough to take his new bride back to New York with him. They spent their first Christmas there and had a fabulous Christmas dinner on board the *Missouri*.

After four years service, he worked for Civil Service as an electrician at Jacksonville and Mayport, FL.

Shimp is retired and enjoys woodworking and gardening. He and Grace have been married 52 years.

ALBERT J. SIEDLARCZYK, born March 14, 1927 in Binghamton, NY, entered the USN March 21, 1944, stationed Sampson, NY; Newport, RI; USS *Wyoming;* USS *Chilton* and Brooklyn, NY. Plankowner on the USS *Missouri,* where he was stationed until discharge May 13, 1946 in Lido Beach, Long Island, NY. He received an AAS ME and AAS EE Degree.

He was a member of the 7th Div. 5"-38 mount and remembers when the burning wing of a Japanese kamikaze bounced off his mount during the Okinawa campaign. He achieved the rank of S1c.

He was awarded the American Theatre, Asiatic-Pacific w/3 stars, Victory, Occupational Service WWII, Philippine Liberation, Philippine Independence and Good Conduct Medals.

Siedlarczyk married Molly Sullivan and they have two daughters, Irene and Marie. He retired from IBM Owego, NY after 25 years, Dec. 31, 1989. Served as USS *Missouri* (BB-63) Association Chaplain four years. Conducted the memorial service on USS *Missouri* at Bremerton, WA, commemorating the 50th anniversary of the Japanese surrender. Presently he is an RSVP defensive driving instructor and a church and community volunteer.

HERBERT D. SIEMENS, born April 3, 1949, North Hollywood, CA, enlisted in the USN April 13, 1967 and was stationed on USS *Monticello* (LSD-35), USS *Beaufort* (ATS-2); USS *Ashtabula* (AO-51); USS *San Jose* (AFS-7); USS *Mount Baker* (AE-34); USS *Santa Barbara* (AE-28); SSC San Diego, CA; NTC Great Lakes, IL; Naval Hospital Charleston. Served aboard USS *Missouri* in Nav. Div. from March 1985 until June 1987.

Memorable experiences include making chief quartermaster; navigating the ship around the world; reenlisting on the surrender deck with his wife and kids.

He will complete 30 years of naval service (over 20 years of active duty) on Sept. 31, 1997 upon retirement from Naval Hospital Charleston.

He was awarded the Navy Commendation Medal, Armed Forces Service Medal, Navy Achievement Medal w/Gold Star, Kuwait Liberation (Saudi Arabia), South West Asia Service w/2 stars, Kuwait Liberation (Kuwait) Medal, Good Conduct Medal w/ 4 Gold Stars, Navy Unit Commendation w/Gold Star, Meritorious Unit Commendation w/2 Gold Stars, Navy "E" Ribbon (2), National Defense w/2 Gold Stars, Armed Forces Expeditionary Medal w/3 Gold Stars, Armed Forces Expeditionary Medal w/3 Gold Stars, Sea Service Deployment Ribbon w/7 stars, Vietnam Service Medal, Republic of Vietnam Campaign Medal w/3 Gold Stars, Republic of Vietnam Gallantry Cross and was surface warfare specialist designator.

Siemens is married to Zamora Preo Fulgosino of the Philippines and has two sons Michael and William.

HARRY SIMMONS JR., born April 13, 1921, Olympia, WA, enlisted in USN Sept. 1, 1942, Seattle, WA. Picked for USS *Missouri* crew at NTS Newport, RI by Cdr. Jacob L. Cooper. He was a yeoman second class and their ship's office was on the main deck. They put in operation ship's personnel records etc. He was at commissioning when the Senator Harry S. Truman and Margaret were there. While in Brooklyn Navy Yard the *Missouri* had been damaged in bow by destroyer. They were towed to Bayonne dry-dock in New Jersey for repairs.

Passing under the Brooklyn Bridge the antennas on super structure had to be lowered to let ship pass beneath bridge. On shakedown cruise he held Captain's Mast east day with then Capt. William M. Callaghan on deck where surrender was signed. Left ship before it left New York for Pacific. Retired 15 years from Puget Power Electric Company.

ANDREW C. SIMPSON, born July 22, 1924 in Washington, DC, joined the USN March 18, 1944 and assigned to the USS *Missouri* June 11, 1944, stationed at deck force, 40mm antiaircraft guns.

His memorable experience includes being part of the 3rd Fleet Naval Landing Force to occupy Yokosuka Naval Base.

He was discharged March 21, 1952 as S1c and awarded the American Theatre w/Asia Clasp, Asiatic-

Pacific w/3 stars, Victory Medal, Japanese Occupational, Good Conduct and National Defense Medal.

Simpson is married to Mary Fay and they have three daughters, Carolyn, Donna and Janice and one granddaughter, Savanna. He has retired from the federal government.

DONALD F. SLACK, born Feb. 21, 1927 in Medford, MA, enlisted in the USN March 1944. Completed high school education through GED in Brooklyn, NY.

He went to NTS Sampson, NY for boot camp, then to Newport, RI. Assigned pre-commissioning detail for USS *Missouri* (BB-63).

Arrived June 1944 and was a plankowner aboard the USS *Missouri* (BB-63) until 1950 after the *Missouri* ran aground in Hampton Roads, VA.

He was first assigned as gun striker in Turret #1 later assigned to 4th Div. for four years, then to FM Div. Ordnance Stores, then to ships armory.

On Sept. 2, 1945 he was escort to correspondents and dignitaries coming to the surrender ceremonies. Most coming aboard the port side and being escorted to surrender deck on starboard side.

He married Margaret Sweeney in 1948 and they have two sons, one daughter, seven grandchildren, one great-granddaughter. He will celebrate his 50th anniversary in December 1998.

Slack retired 20 years USN 1964, started a hydraulic business and sub-contracted for shipyards and retired 1994.

His memories of his naval career and time served aboard the USS M*issouri* (BB-63) have been cherished with the help of the Missouri Association.

EDWIN SMITH (ASIMAKOUPOULOS), born March 14, 1926, and drafted in 1944 and transferred from the 62nd replacement to the USS *Missouri* in the spring of 1945. His duties aboard ship were as gunner on the 20mm gun station by the second turret (port side), search light operator and office clerk for the Marine detachment aboard ship. The *Missouri* was the flag ship for the 58th Task Force, and later the 38th Task Force in the Pacific Theatre, with Adm. "Bull" Halsey in command of the 3rd Fleet. From Leyte, in the Philippines, they bombarded Okinawa, the coast of Honshu, the Hokkaido steel mills, the Kyushu airfield, and the industrial targets of Hitachi, only 60 miles from Tokyo.

On August 9 they fired on their last Japanese plane, a flying Zeke kamikaze, headed for their flagship. It was shot down, thank God!

He was chosen to be a Honor Guard escort to the Russian delegates at the surrender ceremony in Tokyo Bay on Sept. 2, 1945. They were awarded several ribbons and Battle Stars for their action in the Pacific Theatre.

They took part in the Navy Day celebration in New York and were reviewed by President Truman, Secretary of State Forrestal, and Governor Dewey. What a day that was.

He was transferred to the Naval Hospital at St. Albans, NY. After four months he received his discharge in February 1946.

Later he became a mortician and then a protestant minister. He married Star Birkeland in Seattle, WA in 1951. They have two sons, Gregory and Marc and five grandchildren.

In 1968 he changed his last name (Smith) back to his "Greek" father's family name of "Asimakoupoulos".

FULTON B. SMITH, born Sept. 12, 1928 in Milton, MA, joined the USN in September 1948 and was assigned to the USS *Missouri* in January 1949, first in F Div., then in C Div. and received a Letter of Commendation from Capt. Irving Duke for "meritorious performance of duty during operations against the enemy in the Korean Theatre 1950-1951." Discharged August 1952.

Ordained Episcopal priest in 1959 and worked in rehabilitation and social work and parishes. Retired in 1991 and loving it.

PAUL SMITH, born Aug. 14, 1925 in Quincy, IN, graduated from Quincy High School, enlisted in the USN on Jan. 19, 1943. He was stationed at NRS, Great Lakes, IL; USS *New York*; NM, Norva; NTS, Newport, RI; NYD, Brooklyn, NY; USS *Missouri*; USS *Philadelphia*; RS, Pier 92, NY; RS, San Diego, CA; RS, Bremerton, WA and PSC, Lido Beach, Long Island, NY.

He was discharged as an electricians mate third class on Feb. 24, 1946. He received the Asiatic-Pacific Medal w/3 stars, American Theatre Medal and the Victory Medal. Smith was very proud of the time he served on the USS *Missouri* and shared several memories with his family. Before entering the service he married his first wife with whom he had seven children. He later moved back to his hometown and married Mabel L. Myers Neier. They had one daughter and one grandson. He retired with Renners Express due to illness. Smith passed away August 1988 with lung cancer.

JACK E. STEELE JR., born Oct. 21, 1930 in Burlington, NC, joined the USN, June 24, 1948 and was assigned to the USS *Missouri* May 28, 1952, as main radio/radioman.

While at sea he participated in Operation Mainbrace. Started taking water through leak in hull, 75 mph winds and heavy sea, 40 foot waves. His memorable experience occurred during GQ. He did not make it to his battle station. Locked in compartment under No. 2 main battery over two hours while the 16" guns were firing.

Steele was discharged as RM3 Oct. 17, 1952 and received the Unit Commendation, Navy E, Good Conduct, National Defense, Occupation (Europe), United Nations, Navy Expeditionary Medal.

He married Katherine Edwards and they have a daughter, Judy Carol and son, Jack E. III. They also have a grandson, Brandon Michael Steele. He retired as professional engineer, S.C. Electric & Gas in 1989; ham radio and computer guru.

MILTON J. STELTZ, born Oct. 13, 1918 in Chester, PA, joined the USN March 18, 1944 and was assigned to the USS *Missouri* June 1944, 2nd Div., 16" guns as turret striker.

His memorable experiences includes being stationed in *Missouri* and going through the Panama Canal.

He was discharged from the Separation Center in Bainbridge Dec. 16, 1945 as S1c. He married Elizabeth May and they have two sons and a daughter. Steltz retired from Chrysler Corp., Newark, DE.

JOHN (JACK) STEMPICK JR., born Dec. 2, 1931 in East Haven, CT, entered the USN Oct. 11, 1950 and assigned to the USS *Missouri* August 1952. He served as gunners mate third class, trainer operator in turret 2 until March 1944.

His most memorable experience occurred when the order came to train turret out to port and prepare for a "broadside". All turrets and portside five inch mounts fired at the same time against enemy in Korea. Light bulbs broke, anything not secured flew everywhere. He had never shook like that in his life.

He received the National Defense Service, China Service, United Nations Service, Korean Service w/2 stars, Navy Occupation Service, Good Conduct Medals and the Korean Presidential Unit Citation.

Stempick is married to Rachel Maiolo and they have two daughters, three sons and nine grandchildren. He is currently an officer in the USS Missouri Association.

ERNEST STEVENS III, born Feb. 2, 1967 at Whidbey Island NAS, WA, enlisted in USN on June 23, 1985, was assigned to battleship *Missouri* in January 1986 at Long Beach, CA and was among the re-commissioning crew in May 1986 in San Francisco, CA. Served on *Missouri* until June 24, 1989 as a personnelman third class when he was discharged after four years active duty.

The most memorable part of *Big Mo* was the around the world good-will tour in 1986 shortly after re-commissioning. There were many ports of calls from

Australia, Philippines, Persian Gulf, Suez Canal, Turkey, Italy, Portugal through the Panama Canal and back home at Long Beach, CA.

During his service many honors were bestowed upon the ship and its crew, of these he received the Meritorious Unit Commendation, Battle Efficiency Award, Good Conduct Medal and the honor of being a Golden Shellback.

He is currently working as a records manager for The Money Store in Sacramento, CA.

SALVATORE J. STILE, 4th Div., Queens, NY, born Feb. 1, 1933, Brooklyn, NY and was inducted into the USN April 17, 1952. He was assigned to the Deck Force, Aug. 5, 1952.

Memorable experiences: the tragedy of wartime took both an emotional and physical toll on the USS *Missouri* and crew. On one occasion their spotter helicopter was shot down along the Korean coast and three officers aboard were killed. He was stationed on the lookout watch and remembers the sad sight of the planes searching for the officers and the spotter helicopter. This incident has been in his mind. Subsequently, the USS *Missouri* cruise book was dedicated to the memory of those three officers: 1st Lt. Robert Dorman Dern, USNC; 1st Lt. Rex Donald Ellison, USMC; and Ensign Robert Leland Mayhew, USNR.

On another occasion, to the shock and grief of their entire crew, their captain, Capt. Edsall, died of a sudden heart attack while on the bridge of the *Missouri* as it was returning to port in Sasebo, Japan.

One of his fondest memories aboard the USS *Missouri,* was the occasion when he met his brother, Frank, in Korea. The *Missouri* was off port in Pusan Harbor to replenish supplies after a gun strike in Korea. They were young Brooklyn boys. He was a seaman on the *Missouri* and his brother, Frank Stile, was in the 501 Harbor Craft Unit in Pusan, Korea, serving as chief engineer on Gen. Van Fleet's yacht. As the *Missouri* was anchored to a buoy, his brother asked permission from Capt. Frank Wentworth of Massachusetts if he might have time off to see Salvatore as they were both serving in Korea at the same time. Amazingly, Capt. Wentworth responded that not only would he allow his brother Frank to visit with him aboard the *Missouri,* but that he would also transport him to the *Missouri* on the general's yacht. He added that after the on board visit he would return them and they would continue their reunion on board Gen. Van Fleet's yacht and the officer's club. It was about 11:00 a.m. and after radioing the *Missouri* that Gen. Van Fleet's yacht wanted to come alongside the ship, his brother Frank arrived. They toured the *Missouri* and even had chow together. Later, as he waited on liberty call, the side boys were called to duty as Gen Van Fleet's yacht was returning. He will never forget the look of amazement on the face of the officer of the day as his brother and he went down the ladder to board Gen. Van Fleet's yacht to continue his liberty and their reunion. Needless to say, the officer of the day approached him stating that he was not too happy about calling the sideboys to attention.

Stile was honorably discharged in April 1956. His awards include the National Defense Service Medal, Korean Service Medal, China Service Medal and the United Nations Medal.

He is married to Clare Marie Barry, and they

have two sons, Salvatore and Damien, and one daughter, Jillian. Salvatore is currently officer and owner of several corporations and remains active in his business. Salvatore and his wife continue to reside at their home on Long Island, NY.

JOHN P. SULLIVAN, born March 17, 1926 in Falls River, MA, enlisted March 15, 1944. He spent one month in Sampson, NY and one month in Newport, RI. He then went to Brooklyn Navy Yard for precommission duty. Stayed on the *Missouri* until discharged May 13, 1946 in Boston, MA.

He has a lot of memories: the actual signing of peace; a Japanese suicide plane, after which their chief signal man and two first class, made a Japanese flag for burial service at sea. After the war they returned the Turkish Ambassador's body to Turkey.

Sullivan was discharged as S1c signal division and received the American Theatre Medal, Pacific Medal w/3 stars, and the Victory Medal.

He is married to Mary C. Loftus and they have two sons and two daughters: Mark, Kenneth, Claire and Nancy. He is a retired parts manager after 20 years with Cadillac, Oldsmobile, Buick, Pontiac dealer.

CHARLES G. SURBER, born Dec. 21, 1928 in Indianapolis, IN, joined the USN in 1947 and was assigned to the USS *Missouri* in 1947 stationed in the post office.

While at sea he participated in taking midshipmen on training cruises and was involved in military maneuvers.

His memorable experiences: meeting shipmates from other states; visiting foreign ports and having President Truman on board ship.

Surber was discharged in 1950 as TE3c and received the "E" Awards.

He is married to Carol Jean and they have a son, Greg, daughter-in-law Cindy and daughter, Carol Raye.

ROBERT H. SWART, born March 17, 1928 in Boston, MA, joined the USN in 1945 and served on the USS *Tarawa* 1945 to 1947. He served on the USS *Albermarle* (AV-5), 1947-1949 and the USS *Missouri* (BB-63) 1949 to 1952.

He was self-employed for 25 years as a restaurant owner and worked for 23 years as a bus driver. Now retired and works part time doing funeral work. He has a wife, three daughters and five grandchildren and now lives in Weymouth, MA.

GIL TAIMANA, born in Papeete, Tahiti, French Polynesia and joined the USN Nov. 17, 1986 and as-

signed to the USS *Missouri* February 1987 as electrician/safety shop/switchboard.

While at sea he served as escort of Kuwaiti reflag tankers through Gulf of Hormuz.

His memorable experiences include sailing into Sydney Harbor with a flotilla of small boats around them.

He was discharged as PO3/EM3 November 1989 and received the Sea Service Medal, Battle E, Expeditionary Forces and Navy Commendation. Taimana is now a graphic artist and computer web publisher.

DANTE (DAN) L. TANCREDI, born March 10, 1917 in Boston, MA, entered the USN May 3, 1943. Stationed at Newport, RI, USS *New York* and Brooklyn, NY. Plankowner on the USS *Missouri.*

His memorable experiences include repairing Adm. Halsey's shoes and receiving a $5.00 bill signed by Adm. Halsey for a job well done. Do not ask him what he did with it.

He was discharged October 1945 from Norfolk, VA and was member of the SS Div. 20mm antiaircraft. He remembers Okinawa, Iwo Jima and the bombardment in Hokkaido, Japan.

His awards include the American Theatre, Asiatic-Pacific w/3 stars and the Good Conduct Ribbon.

Tancredi is a retired shoe worker and lives with his wife Lucy and enjoys his four grandsons: Michael, Carl, Paul and John.

GEORGE THEBERGE, born March 22, 1932 in Haverhill, MA, joined the USN May 25, 1950 and went aboard the USS *Missouri* after boot camp around August 1950. He was aboard for both tours to Korea, and left the *Missouri* in May 1953. He transferred to submarine base, New London, CT and his daughter, Belinda was born at the base hospital in 1954.

He was discharged in May 1955 and is a retired iron worker.

SCOTT A. THORNBLOOM, born June 19, 1962 in Port Byron, IL, enlisted in the USN Nov. 2, 1980. His duty stations included RTC/NSHS, San Diego, Great Lakes, IL, Regional Medical Center, 1st Marine Div., Camp Pendleton, CA, Indianapolis, Naples, Italy, USS *Missouri,* NBS/FSD, San Diego, PS&S, Tokyo.

Thornbloom was a member of X-3 Div. which later became *Mighty Mo's* Public Affairs Department. He served aboard BB-63 as a journalist/broadcaster from January 1989 to January 1992. His memories of *Missouri* are woven through the many stories he wrote about the ship and her crew. He remembers how serene and majestic she looked each morning moored to the pier. He reflects on how huge the ship was when standing underneath her at the bottom of the Long Beach Naval Shipyards drydocks. He recalls Cher, SINKEX 89 and Christmas in Mazatlan, Mexico. And he'll always remember how well they came together during Desert Shield/Desert Storm. He remembers the "night of the gas attack" finding the CIWs round an an O-2 level stateroom passageway and working in turret #2 as a projectile hoist operator during GQ. He'll never forget the rousing cheers the ship and crew received after the Gulf War when it arrived first in Freemantle/Perth, Australia, then in Hawaii and finally back in Long Beach, CA.

His awards include the Joint Achievement Medal, Army Achievement Medal, Combat Action Ribbon, Navy Unit Commendation, Meritorious Commendation, Battle E x3, Good Conduct Medal, Fleet Marine Force Ribbon, National Defense, Southwest Asia, Sea Service, Overseas Tour x 3, Saudia Arabia Medal, Kuwaiti Medal and .45 pistol.

Today he is a journalist first class petty officer at the CINCPACFLT Public Affairs Office in Pearl Harbor and is anxiously awaiting *Mighty Mo's* arrival in 1998.

MARIO (MARTY) THUMUDO, born April 30, 1932 in Fairhaven, MA, joined the USN Sept. 1, 1950 and was assigned to the USS *Missouri* May 1951 stationed in band.

While at sea he participated in Korea October 1952-April 1953. Every day of his enlistment was a memorable experience.

Thumudo was discharged as Mu2 June 30, 1954 and received seven awards. He is now retired and has three children and four grandchildren.

ALFRED L. TICEHURST, born Dec. 9, 1928 in Burlington, VT, joined the USN Jan. 9, 1946. He was stationed with the 7th Div. and achieved the rank of seaman second V6 USNR.

His memorable experiences include going to Turkey to bring back the body of the Turkish Ambassador.

He was discharged Sept. 26, 1946 at Bainbridge, MD and received the Victory Medal.

On March 28 he married Theresa Rocheleau and they have two children, Richard and Karen, three grandchildren and one great-grandchild. He is now a retired teamster.

ROBERT G. (BOB) TOMANEK, born Sept. 3, 1929 in Binghampton, NY to Frank and Rose. Joined the Boy Scouts at 11, Sea Scouts at 15, NY State Guard at 17, USNR at 18, regular Navy at 19 in 1948. Left the *Mo* five days before she ran aground in 1950 for machinery repair school, San Diego. Graduated top of the class. He was on the *Missouri* during its Korean conflict, leaving it on its return to Norfolk for a two month leave. Reassigned to the USS *Salem* operating in the Mediterranean. Flew from Malta to Boston for discharge Sept. 4, 1952 as MR2 with Naval Occupation (Asia), Korean Service (2 stars) and UN Medal.

One experience: in early 1949 on a cruise to port of Spain, Trinidad he had liberty boat duty. Towards the end of the day, coming up to the boat boom to tie up, the current was kind of strong. He held the boat for the other crew members to climb the ladder. When his turn came he just about grabbed the bottom rung. After hanging on for as long as he could, he fell in the drink. He never did thank the guy in the other boat who pulled him out. All he could think of at the time, was when are the Barracuda going to hit.

He married Lena Masaryk in 1955 and they had four wonderful kids: Bob Jr.; Paul with two grandchildren, Gina and Paul Jr.; LeAnn and Midge. Retired from IBM in 1987 as senior design automation specialist. His drawing was printed for the ships *Domain of The Golden Dragon*.

DANIEL TOOMEY, born April 12, 1926 in Limerick Ireland, enlisted in the USN March 17, 1944 (St. Pats Day). Assigned to USS *Missouri* June 1944 and served aboard as lookout and S1c 3rd Div. Saw action at Iwo Jima, Okinawa and bombardment of Japan. He was discharged May 13, 1946 as S1c.

Most memorable experience: Battle Station Sky Forward (from Crows Nest) birdseye view of all the action, some good, some bad.; Jap planes being shot down, some pilots close enough to see their face; being hit by kamikaze; carriers also being crashed on by kamikazes; their planes coming back from raids disabled and trying to make it to their carriers, sometimes plunging in the ocean or crashing on the carrier.

Today he is a retired civil engineer. He and his wife Jane had seven children: Kevin, Daniel, James, Brian, Kathleen, Patrick and Thomas and 11 grandchildren.

CYRUS TOPOL, born Oct. 2, 1911 in Boston, MA, enlisted into the USN March 1942. Attended basic training in Newport, NTS; Gunners Mates School, Great Lakes, NTS as student and instructor for two years; Advanced GM Electric and Hydraulic School, Washington, DC Navy Yard 1944. Assigned to USS *Missouri* May 1944 and sent to Brooklyn Navy Yard to observe construction of gun mount (5"38) as GM2c. He was on board for commissioning June 1944, serving as GM1c in charge of gun mount #3 until end of war and was honorably discharged in October 1945 at Boston.

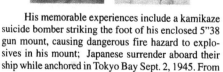

His memorable experiences include a kamikaze suicide bomber striking the foot of his enclosed 5"38 gun mount, causing dangerous fire hazard to explosives in his mount; Japanese surrender aboard their ship while anchored in Tokyo Bay Sept. 2, 1945. From

his perch standing on top of his mount he was an eyewitness to the entire proceedings from 50 feet behind MacArthur's neck from beginning to end.

DEMETRIUS PETER TRAGGIS, born Jan. 22, 1926 in New Haven, CT, was inducted into the USN March 17, 1944 and assigned to the USS *Missouri* May 5, 1944, CR Div., GQ gunner on 40mm station.

His memorable experience includes all hands working together and doing their job; waiting on Adm. Halsey when he was doing his duty in charge of chiefs dining room, during his three months duty.

He was awarded the American Theatre Medal, Asiatic-Pacific Medal w/3 stars and the Victory Medal and discharged May 13, 1946.

His father served in WWII. Traggis has been self-employed. He is now retired. Witnessing the surrender had a big effect on him. He's still trying to do his part in saving our country from complete financial disaster.

LEWIS F. (FRANK) TRAVIS, born March 8, 1932 in Keene, NJ, enlisted in USN Aug. 22, 1950, attended boot camp and Class A ET School at Great Lakes 1950-1951. Boarded the *Missouri* in Norfolk in 1951 and left her in 1954 for duty at mothballed carrier fleet in Boston. He was discharged June 18, 1954.

Memories include being aboard the *Missouri* on her second Korean cruise, where they sadly lost their spotting chopper and crew, as well as their captain. More pleasant times were had on their middie cruises.

Travis was discharged an ET3 and received the National Defense Service Medal, Good Conduct Medal, Korean Service Medal w/2 stars, UN Service Medal and the China Service Medal.

He is married to Martha Grace and they have four children: Mark, Michael, Manapa, Rebecca and four grandchildren: Katie, Ben, Meaghan and Leanna.

He received a BSEE from UNH (1958) and MSEE from Northeastern (1961); retired from Bell Telephone Laboratories after over 35 years in product design on June 31, 1993.

KENNETH R. TRIGG, born Feb. 17, 1934 in Peoria, IL, joined the USN September 1941 and was assigned to the USS *Missouri* December 1951 and stationed in the #2 Fireroom and achieved the rank of BT3.

While at sea he participated in the Korean campaign 1952-1953. His memorable experiences include getting off watch and finding all lights lit, est. 400 men reported to sick bay with diarrhea and another 400 who didn't. Afterward there was no bucket to be found.

He was discharged Dec. 29, 1954 as BT3 and received a shellback certificate.

Trigg has been married twice and has three children by his first wife and nine grandchildren. He is a retired boilerman and enjoys rebuilding and restoring farm tractors, playing pool and golf.

WILLIAM J. TULLY, born Nov. 1, 1926, entered the USN Nov. 1, 1942.

His memorable experience: 8th Div.; always had a cigar going; once while firing on the guns (20mm) starboard side, one mount was changing barrels, one of the crew forgot to lock barrel of the (20mm) and it flew overboard.

He was awarded the Asiatic-Pacific Area w/3 stars and discharged January 1946.

Tully married Lillian Dooley May 18, 1946 and they have children, Bruce and Diane; grandchildren, Elizabeth and William; a great-grandchild, Shelley. He has retired from the private sector.

PAUL (BUD) VACCARI,

born Feb. 10, 1924 in Hartford, CT, entered the USN December 1943 and was assigned to the USS *Missouri* June 1944. He was a WT3c in fireroom #2.

The action he saw at sea was at Okinawa, Iwo Jima and the bombardment of Maman Iron Works in N. Hokkaido, Japan.

He remembers the heavy smoke pouring from the fresh air vents the day the kamikaze plane hit them and not knowing what was going on.

Vaccari was discharged Jan. 18, 1946 with three Pacific Stars. He has four sons, a daughter and three grandchildren. He and his wife Rita travel often, visiting two of their sons who are Navy Seals.

RICHARD D. VAIL,

born Sept. 26, 1926 in Long Beach, CA, joined the USN Aug. 30, 1944 and was assigned to the USS *Missouri* Dec. 18, 1944 in San Francisco, CA. He stayed aboard the *Missouri* through all the big Pacific battles and entered Tokyo Bay for the surrendering signing. After the return to New York, he signed up for the trip to Europe which included Turkey, Greece, Capri, Algiers and Tangiers.

Vail was discharged June 18, 1946 and returned to his home in Ojai, CA, married his high school sweetheart, Shirley and they had three sons: Sam, Mike and John. He spent 40 years working in Citrus Management Farming. He died Sept. 23, 1995 and was buried on his 69th birthday, Sept. 26, 1995.

FRANK VASQUEZ,

born July 14, 1925, enlisted in the USN July 13, 1943. Attended basic training at Great Lakes NTS, Chicago, IL. After 10-day leave he boarded troop train for Pleasanton, CA, Sept. 10, 1943. Left Treasure Island aboard troop transport USS *Sea Pike* Sept. 28, 1943 and arrived Brisbane Australia, Oct. 23, 1943. He boarded USS *General John Pope* and arrive in Milne Bay, New Guinea, USN Supply Depot 1967. Spent 27 months in New Guinea and achieved the rate of S1c to ships cook second class petty officer in charge of watch. At 18 years of age he was in charge of at least 18 cooks and working parties. He left New Guinea to Manus Island, Admiralty Island chain. Boarded SS *Franz Sigel* Sept. 30, 1945 and arrived in the Panama Canal Oct. 30, 1945. He went to Galveston, TX Nov. 5, 1945. After a 30 day

leave he reported to Brooklyn Navy Yard, Dec. 13, 1945 and was assigned duty aboard the USS *Missouri* Jan. 16, 1946. Served on the *Missouri* until discharged at Lido Beach, Long Island, NY, May 13, 1946.

While aboard the *Missouri* he served in commissary as second class petty officer in charge of starboard watch. He crossed the Atlantic several times, visiting the following ports: Rock of Gibraltar, Algiers, Tangiers, Naples, Italy, Pireius, Greece and Istanbul, Turkey.

The *Missouri* was the first American Navy ship to enter Turkey since 1939. Turkey was a neutral country during WWII. Schools and factories were closed for three days while they were in port. The USS *Missouri* was accepting visitors during their stay and the lines were very long. While in Naples, Italy they were given a three day leave to Rome. At that time the Air Force put them up for three days at Mussolini's headquarters in Rome.

Vasquez received the Victory Medal, American Theatre Medal and the Asiatic-Pacific Medal.

He is married to Doris Hunt and they have a daughter, Sue and son, David. He is retired from American National Can Company as production mechanic.

JOSEPH S. VELLA,

born April 17, 1926 in Thompsonville, CT, enlisted in the USN at 17 and was aboard for the commissioning of the USS *Missouri*. He was discharged in 1946 at Lido Beach, NY, member of 6th Div., 5"38, Mount 10.

Vella remembers the day the word came over the PA system that the Japs surrendered. He was standing with an old salt named Mac McDonald, who had a ship blown out from under him. He grabbed Vella by the shoulders crying and said, "Joe, I made it," with tears in his eyes, Vella said, "Me too".

He received the American Theatre Medal, Asiatic-Pacific w/three stars, Occupational Service, WWII Victory, Philippine Liberation, Philippine Independence and the Navy E. Ribbon.

Vella is a retired carpenter after 35 years. He married Christine Carenza and they have two daughters and one son. He is one of the founders of the USS Missouri Association, member of the Color Guard and has held several positions in the association and is presently historian.

JOSEPH P. (BUDDY) VERDI,

born Aug. 31, 1925 in Brooklyn, NY, joined the USN December 1942 and was assigned to the USS *Missouri* November 1943 in combat info center.

His memorable experience: he trained the whole division at Fighter Dir. School and radar operation. Chauffered Cooper the whole time the ship was in Brooklyn. He was discharged June 1945 and received the American, European and Asiatic Theatre of Combat and Victory Medal.

Verdi is now divorced and has two children, seven grandchildren and one great-grandchild on the way. He is a Big Band Leader, real estate and mortgage broker.

OSWALD WILLIAM VOGEL,

born Aug. 29, 1910 in Lawrence, MA, joined the USN Jan. 17, 1944. He had been a professional musician for 15 years and retired after 23 years as band director and teacher from the Lawrence, MA School System.

After basic training and radar schooling he was assigned to the USS *Missouri* and witnessed the commissioning. He was on the shakedown cruise and

eventually was sent to the battle area in the Pacific. He was assigned to be a "side boy" at the signing of the surrender and saw the Japanese dignitaries at close hand. It was a momentous occasion; one he shall never forget.

He was awarded the American Theatre, Asiatic-Pacific w/3 stars and Victory Medal.

Vogel has been happily married to his wife Marianne for 60 years and they have three daughters, ten grandchildren and ten great-grandchildren.

WILLIAM T. WALSH JR.,

born Jan. 28, 1927 in Hartford, CT, enlisted in USN March 10, 1944. He was aboard the USS *Missouri* when it was commissioned in Brooklyn, NY. He was in the 3rd Div. and received the American Campaign Medal, Asiatic-Pacific Campaign Medal, WWII Victory Medal, Navy Occupation Service Medal w/Asia Clasp. He was SFC when he was discharged on May 13, 1946.

His most memorable event was when he witnessed the signing of the Japanese surrender aboard the ship and his picture was in *Life Magazine*. He was one of the founders of the USS Missouri Association and was also a plankowner.

Walsh was married 45 wonderful years to Terry Walsh and they had four sons, one daughter and 12 grandchildren. He was employed by the SNE Telephone Company for 35 years and a volunteer fire chief in Windsor, CT for four years. He passed away Aug. 5, 1995.

JOHN JOSEPH WARNER,

enlisted in the USN Dec. 13, 1946 and belonged to the Roger Div. USN Dec. 13, 1946 to Nov. 6, 1951. He was stationed at Anacostia Air Station, Washington DC, April 9, 1947 to April 28, 1949.

He boarded USS *Macon* (CA-132) May 8, 1949 to April 1950. USS *Macon* was commissioned Aug. 26, 1945 and decommissioned April 14, 1950.

Warner was aboard USS *Missouri* (BB-63) April 18, 1950 until discharge Nov. 6, 1951. Rated an extra year courtesy of President Truman.

He was awarded the Korean Service Medal and the Navy Occupation Service Medal. He served in the USN four years, ten months and 24 days. Of that time, two years, six months was served overseas.

He worked all his life as sheet metal worker, a trade he started to learn in the USN. He passed away Aug. 11, 1990 from lung cancer. At the time of his death, he was survived by his wife of 25 years Katherine, and his son, William.

ZANE THOMAS WATTS,

born Nov. 5, 1961 in Winslow, AZ, entered the USN in March 1982.

His stations include: USS *Enterprise*, NMRC Phoenix, USS *Missouri* and SIMA NRMF Long Beach. Served on board USS *Missouri* March 1988 and decommissioned in 1992.

He was a member of F Div., Forward Main Battery Plot (16" gunfire control) as a stable vertical operator. He remembers 37 16" gunfire mission in Kuwait. Hands sweating on 16" firing keys made of brass while turning them green from oxidation, from 58 straight hours of gunfire missions during the 100 hour offensive.

Watts received the Navy Achievement Medal, Combat Action Ribbon, Navy Unit Commendation and other unit awards.

He is a proud father of a beautiful girl named Dana. He works as a electromechanical technician in the area of Materials Science with Instron in Canton, MA.

EUGENE O. WEBBER, born May 16, 1923 in Worcester, MA, entered the USN April 1942 in Worcester, MA and was sent to Great Lakes Training Center.

August 1942 he joined the USS *Omaha* in the south Atlantic and made coxswain rate. Put in for transfer to USS *Missouri* and flew to NYC for the commissioning.

While in the 7th Div. the most exciting event he experienced was during the surrender signing when he escorted a journalist, who did sketches, to the main battery director where he could look down on the ceremonies. As they viewed the deck below they noticed another fellow filming everything around them. The reporter made a remark that Webber will never forget concerning the Russian filming nearby. "He may be filming these ceremonies with the Japanese, but our real enemy is right there!"

He was discharged November 1945 as coxswain. He married in 1947 and has seven children. Retired after being in child evangelism fellowship for 30 years. They attend reunions when possible.

ROBERT J. WEBBER, born July 9, 1957 in Evanston, IL. After graduating in June 1975 from Cary-Grove High School in Cary, IL, enlisted in the USN on Sept. 23, 1975. Stationed on USS *Kansas City* (AOR-3), USS *L.Y. Spear* (AS-36), USS *Sacramento* (AOE-1), NTC Great Lakes, IL and Long Beach, CA where USS *Missouri* (BB-63) was outfitted for recommissioning in July 1985. Served aboard *Missouri* in E and X-4 divisions until decommissioning on March 31, 1992.

"Spending one-third of my naval career aboard the *Missouri* was an honor he will never forget. So much has happened while he served on board her. They managed to dodge a scud missile attack in the Arabian Gulf off shore of Kuwait in February 1991, while he was at general quarters in No. 4 Engineroom on the switchboard. He will always remember, while in Pearl

Harbor, HI the time he shook hands with then President George Bush onboard *Missouri* during the 50th anniversary of Pearl Harbor, Dec. 7, 1991. A world cruise, two tours to the Arabian Gulf including Desert Shield and Desert Storm, and many other exotic ports. The power of the 16-inch guns has never been matched. He was sorry to see her leave the fleet."

Completed more than 20 years of active naval service at Naval Training Center, Great Lakes, IL on May 31, 1996 as a senior electrician's mate. Awarded Surface Warfare Specialist Designator, Navy Achievement Medal w/Gold Star, Combat Action Ribbon, Navy Unit Commendation, Meritorious Unit Commendation w/Star, Navy E Ribbon, Navy Expeditionary Medal, National Defense Service Medal, Armed Forces Expeditionary Medal, Southwest Asia Service Medal w/2 stars, Sea Service Deployment Ribbon w/6 stars, Kuwait Liberation Medal (Saudia Arabia) and Kuwait Liberation Medal (Kuwait).

He is married to Edwina Sison Loyola, has three daughters: Jennifer, Maria and Melissa. Currently employed at Barat College of the Sacred Heart, Lake Forest, IL.

GEORGE A. WEIGOLD, born Dec. 4, 1925 in Roslindale, MA, was inducted into the USN June 1943 and assigned to the USS *Missouri* March 1944, stationed on deck crew 20mm antiaircraft.

His stations include: Newport, RI; USS *Tattnal;* Fleet Service School, Norfolk; plank owner USS *Missouri.*

His memorable experiences: as a member of the Antiaircraft Div., he witnessed and actively participated in the battles and events culminating in the surrender of the Imperial Japanese forces in Tokyo Bay.

Weigold received the American Theatre, Asiatic-Pacific w/3 stars, Victory, Occupation Service WWII, Philippine Liberation, Philippine Independence. He was discharged Dec. 7, 1945.

He married Catherine Rokes and is retired after 32 years as officer manager of a major electronics company. His hobbies include: woodworking, gardening, sports buff. Currently serving as public relations officer for USS Missouri Association.

HARRY WEISS, born Aug. 7, 1926 in Saint Nazaire France, entered WWII April 1944, boot camp Farragut Idaho "six weeks". Sent to fight Japs from the Aleutian Isle to the South Pacific aboard the USS *Salinas* (AO-19). Discharged February 1946. Reentered the USN again December 1948 as seaman first class aboard USS *Missouri* 1st Div., No. 1 16" turret, No. 1 barrel primerman. Transferred to engineering, A Div. Went to Korea under the UN flag to flight North Korean and Chinese Communism aggression 15 engagements. Transferred to heavy cruiser USS *Macon* (CA-132) until his discharge October 1952.

He held rates as apprentice seaman, S2c, S1c, Fireman 1/c, engineman third class, engineman second class.

Weiss received the Good Conduct Medal, China Service, American Campaign, Asiatic-Pacific Cam-

paign w/2 stars, WWII Victory, Occupation Service (Asia), National Defense, United Nations, Korean Defense w/four stars, Navy E, Syngman Rhee, Presidential Citation (Korea), President Boris Yeltsin Russian Government Commemorative Medal and from his shipmates - worlds biggest Mo Turkey. Member of USS Missouri (BB-63) Association, American Battleship Association, United States Cruiser Association, United States Cruiser Association and American Legion Post 217 Warsaw Missouri where he lives in retirement.

JIM WENGER, born Jan. 16, 1926 in Rittman, OH, entered the USN June 28, 1944 and was discharged June 4, 1946 as an RM3c.

His discharge lists service on these ships: *Missouri, South Dakota, Quincy, Rocky Mount, Miami, Massachusetts.*

He was called to the 3rdFlt Flag from Aiea, Oahu.

Experiences include his first taste of champagne at Adm. Halsey's residence on Oahu while taking liquor to the USS *Louisville* for transportation to Guam. He boarded the *Missouri* at Guam on May 18, 1945.

The "surrender' ceremony was the big highlight of all time. He remembers the sky opening up to a beautiful Sunday am. The "flag" personnel did not stand quarters resulting in his witnessing the ceremony from an excellent position.

He and Barbara have two sons, Stephen and David, one grandchild and two step-grandchildren.

Wenger spent 38 years in Kentucky as mostly an auto parts business owner. Presently retired near Hendersonville, NC where his wife was reared.

LEWIS M. WETZEL, born in Summit Hill, PA on June 30, 1926. joined the USN on March 6, 1944, took boot training at Sampson, NY, Unit G and went to Gunnery School in Newport, RI. He was assigned to the USS *Missouri* on June 11, 1944. His rank was SSML3c in the officers laundry issue room.

He was in three battles earning three stars. Most memorable experience was the USS *Missouri* going into Tokyo Bay and witnessing the signing of Japan ending the war. Also saw President Truman, Adm. Nimitz, Adm. Halsey, Gen. MacArthur, and all the high dignitaries during the surrender signing.

Lewis was discharged on May 12, 1946 at Bainbridge, MD. Today he is retired and enjoys woodworking. He and his wife, Geraldine have been married 51 years and have two sons, Lewis and Jerry, three daughters, Linda, Donna and Melissa, 13 grandchildren and eight great-grandchildren.

PAUL R. WICKENHEISER, born March 10, 1927 in Columbia, PA, enlisted in the USN, March 16, 1944. Attended boot camp at Sampson, NY, plankowner on USS *Missouri* until discharged May 15, 1946 in Bainbridge, MD as a seaman first class.

Memorable experiences: he was a baker and he really enjoyed his work. During long deployments, the flour occasionally had "bug" in it. They told the crew it was "Rye bread". When they made cherry pies, they kept the juice to make wine. He remembers when a kamikaze plane went by the gun he was manning, 15 ft. in front of them. The plane hit their ship and the pilot was cut in half. During the time that the peace treaty was being signed, the sky was full of planes as

far as you could see. What a great day.

He was awarded the Pacific Theatre Ribbon w/ 3 stars, American Theatre Ribbon and the Victory Medal.

He is married to Pauline Bowman and has one daughter and two sons. His oldest son served in the USAF; youngest son currently serves as a major in the US Army. Wickenheiser is retired from Alumax Aluminum Co. and resides in Lancaster, PA.

DENNY W. WILBURN, born April 2, 1932 in

Windfall, IN, graduated from Tipton, IN high school in 1950. Enlisted in USN March 5, 1951 and went through boot camp at Great Lakes. Attended Electronic Technicians School from July 30, 1951 to April 18, 1952. Assigned to the USS *Missouri* in Norfolk,

VA (T Tare Division).

He was a member of the USS *Missouri* basketball team which was a very successful. They won the 7th Fleet Championship. He was involved in the decommissioning of the *Missouri* in 1955 in Bremerton, WA. The last few days aboard ship were very sad. His division shipmates had all been transferred or discharged. All the radio and radar had been "powered down" and the *Mighty Mo* was quiet. He was discharged from Fort Lewis on Feb. 25, 1955 a ET2c. He received seven service ribbons.

He is married to Roberta (Gerstner) Wilburn and has three children: son, Timothy and daughters, Cynthia Dillard and Jennifer Kiner; three grandsons, Jeremy and Joshua Dillard and Tyler Wilburn. He is retired in 1991 from AT&T and Ameritech after 38 years.

WALTER L. (LEE) WILLIAMS, born Sept.

28, 1924, entered the USN April 3, 1943 and was assigned to the *Missouri* May 1944 as Mus3c. The band was assigned to the repair division but transferred to the lookout division when an admiral had to wait an hour to come aboard while the band cleaned up and got into proper uniform. He was selected for flight training and left the ship, by breeches buoy with two others in April 1945, to board the USS *Niobrara* and ship-hitch home by way of Okinawa, Ulithi, Kamikaze attacks. Life on an army transport and finally boarded the USS *Minneapolis* at Pearl and arrived at Bremerton two days before VE Day. Sent to Monmouth College, then Muhlenberg. Discharged from Pensacola April 1946.

He attended the University of Kansas on the GI Bill, received a degree in architectural engineering. Worked in the fabrication and architectural business. Retired in 1986. He and his wife Jackie have two sons, two daughters and eight grandchildren. They live at a residential lake (Waukomis) in Platte County north of Kansas City, MO. Williams enjoys fishing

and golf.

RICHARD WILSON, born June 25, 1918 in

Wilkinsburg, PA, joined the USN February 1944 and assigned to the USS *Missouri* March 1944 as S1c. He was discharged Dec. 25, 1945 as S1c.

He was awarded three ribbons, Asiatic-Pacific Medal w/3 stars, American Theatre Medal and the Victory Medal.

Wilson married Marie and they had sons: Tom, Skip, Gary, Terry and Bob and daughters, Lorraine and Judy. He passed away of cancer Nov. 3, 1996.

JOSEPH G. WIRTH, born Aug. 13, 1926 in Bos-

ton, MA, enlisted February 1944. Trained at Sampson Training Station, NY. Stationed Newport, RI; Brookport, NY. Plankowner on USS *Missouri* until discharge in 1946. Served in 2nd Div. Remembered often the kamikaze plane crashing near the gun mounts.

He did not witness the signing of the peace treaty, as he was with the naval landing force at Hokosuka Naval Base on sentry duty, before the entry into Tokyo Bay by the 3rd Fleet.

Joined the Boston Fire Department after working as a plumber. Retired in 1963 after a severe heart attack in 1962. One of the founders of USS Missouri

Association. Retired to Spring Hill, FL in 1986.

He married Elizabeth Doyle and is the proud father of 12 children. His oldest son was killed in Vietnam in 1970. On April 5, 1992 Wirth died after a third heart attack. He left 22 grandchildren and four great-grandchildren.

EDWARD JAY WOLFE, born in West Hartford,

CT 1926, enlisted in the USN March 7, 1944. Assigned to Brooklyn Navy Yard while USS *Missouri* was being constructed. Served on the *Missouri* from commissioning until discharge May 13, 1946. On board for signing of surrender in Tokyo Bay. Signed on for extra tour of duty following surrender when *Missouri*

took the body Turkish Ambassador back to Istanbul.

Served in L Div., as lookout. At time of discharge was yeoman apprentice. Involved in battles of Iwo Jima, Okinawa, Kyushu and earned the Victory Medal, Asiatic-Pacific Medal w/3 stars, American Theatre Medal.

Memories: being at battle station in #2 turret powder room when kamikaze hit starboard side; ship turned severely and they lost footing and wondered if they were going to get out; signing of surrender while hundreds of planes flew overhead. Seeing bodies of Japanese sailors float by ship in Tokyo Bay.

He married Barbara Hall October 1957. Retired from Avco-Lycoming March 1988. Moved to upstate New York and ran a Bed & Breakfast (Wolfe Hall) for six years. Returned to Connecticut 1994. Now resides in Hamden and has one daughter and one grandson.

WILLIAM W. WOOD, his first glimpse of *Mighty*

Mo was in January 1944, at the Brooklyn Navy Yard where she was still in dry dock. She would be home to him for over two years.

He was born in Protem, MO Aug. 15, 1924, inducted into the USN July 1943 at Kansas City, KS. After boot camp, Electrical School at Purdue University, he graduated as EM3 on December 1943, and trained aboard USS *New York* at Norfolk, VA. After shakedown cruise to Trinidad and return for repairs they left New York for Panama Canal November 1944. Their first casualty occurred in Balboa when a gunnery crew seaman slept in a turret. When the turret was accidentally energized he was decimated. They continued to San Francisco November 1944, then in December 1944

for Pearl Harbor, then to Ulithi Island.

His work station was the I.C. room aft. His battle station was on the superstructure below the bridge. He had a ringside seat when they had their first kill. Antiaircraft guns shot down a Jap Betty. And, off Okinawa, when a crippled Jap Zero crashed into their starboard side, wing aflame, scattering pieces on deck. There was a little damage to their ship.

His most memorable event took place on Aug. 14, 1945 when he received word while on duty in Central Station that the Japanese had surrendered. In ship's log book he wrote "Dirty Damned Japs Surrendered 11 a.m" Best birthday present he ever got. He stood on the superstructure and watched Japanese officials, General Douglas MacArthur, etc. sign the surrender.

In October, 1945 the USS *Missouri* fired 21-gun salute as President Harry Truman came aboard and shook hands with every man from *Missouri*. He remained on board (to Turkey, Italy, Gibraltar, Tangiers, Greece) until his discharge May, 1946.

He received the American Area, Victory Medal, Asiatic-Pacific Ribbon w/3 stars. He is now retired from Sieman-Allis and resides in St. Francis, WI. He and his wife Arlynn have seven children, 13 grandchildren and one great-grandchild.

JERRY WOODS, it all started in early winter of

1944 as SLV Samson Naval Training RI for firefighting school Virginia for gunnery training. His

assignment the *Big Mo* (BB-63).

Radar would be his field. Each day brought new excitement, new adventure and new dangers. They took these all in stride. Once while on the open bridge with fierce seas breaking over the bow with Mom's Kodak pocket camera and with precision timing his shutter matched the perfect wave breaking over the bow of the *Big Mo*. He would like the record to show he took this one time picture which is used on all stationery. They all share wild experiences like failing to hear warnings of a gunnery practice and wound up bounding on the deck. After their share of battles and the signing in Tokyo Bay they left for the USA. Another treat was in store for those who stayed aboard. A tour to the Mediterranean to visit Old World Countries that would provide them with everlasting memories.

OLIVER PERRY WOODWARD JR., born

Feb. 6, 1925 in Allendale, SC, joined the USN Feb. 5, 1943 and was assigned to the USS *Missouri* April

1944, AV Div. Parachute rigger and storekeeper.

His memorable experiences include almost being left being when the fleet left Leyte to bombard Japan and Gen. MacArthur came to his aid when Gen. Southerland started to chew him out.

While at sea he participated in action at Iwo Jima, Okinawa and the bombardment of Japan. Discharged Feb. 20, 1946 as PR2c. He received the American Theatre Medal, Asiatic-Theatre and Victory Medal. Qualified for Good Conduct Medal but never received it.

He is married to Rhidona Reese and they have a son, daughter and three grandchildren. His parents are deceased. Woodward has been employed in general office and warehouse work for the Nu-Idea School Supply Company for the past 45 years.

ROBERT E. WOOLSEY, born Sept. 2, 1925 in Denver, CO, entered USNR Dec. 18, 1943, attended Fire Control School in San Diego. He went aboard USS *Missouri* December 1944 and served until discharge

May 16, 1945 in San Pedro, CA.

He was a member of FA Div., FC3c, assigned to Optical Shop. Duty station: auxiliary director outside 5" director No. 3, just above where kamikaze wing came to rest.

His memorable experiences include a typhoon off Okinawa and damage to 16" turret #1; the Mediterranean tour in 1946; the 50th Anniversary in Bremerton, and Stan Robson.

Woolsey received the Asiatic-Pacific w/3 stars, American Theatre, Victory and Good Conduct Ribbons. He is married to Mary Grannan and they have two sons and one daughter.

He received a BS in education and an MA in Education Administration. Taught elementary school for 39 years in San Fernando Valley, Los Angeles Unified School District. Now enjoying retirement.

CHESTER S. WROBEL, born Nov. 19, 1926 in New Britain, CT, joined the USN January 1944 and was assigned to the USS *Missouri* in 1944 when it was commissioned. He served as gunners's mate on turret #3.

While at sea he participated in the Pacific Area Campaign and was discharged May 1946 as S1c.

Wrobel married Helen Drobot and they have one daughter, Patricia Ferris and one son, James T. Wrobel and three granddaughters. He retired from the VA Medical Center, Newington, CT after 38 years in the Engineering Department.

ANTHONY J. YANNOTTI, born Nov. 15, 1925 in Schenectady, NY drafted into USN during his senior year of high school, March 1944 and later returned to complete his education after the war. Basic training was at Sampson Naval Training Center in Geneva, NY. After boot leave he was assigned to the USS *Missouri* detachment for sea training in Newport, RI. He was present at time of USS *Missouri* commissioning at

Brooklyn Navy Yard on June 9, 1944 which allows him to be plankowner. On board the *Missouri* he was assigned to turret #1, 1st Div. and his job was located in the magazine for the 16" guns. Later he was made "pointer" for the 16" guns.

The USS *Missouri* was bound for action in the Pacific Theatre and joined the 3rd Fleet. The *Missouri* saw action at Iwo Jima, Okinawa and bombardment of coastal installations of Japan.

His most memorable experience was to be present at the formal surrender ceremonies with Japan on the USS *Missouri,* Sept. 2, 1945, and to observe the dignitaries from his assigned position top side. He will never forget the great contingent of allied war planes flying over head at the end of the ceremonies and the pride one felt to say, "Yes I am an American."

Yannotti was discharged from the USN in March 1946, Lido Beach, Long Island, NY as S1c. He received the following medals: American Theatre, Asiatic-Pacific w/3 Battle Stars and the Victory Medal.

He is married to the former Eleanor Pasquarello and they have three sons: Arthur, Richard and James and four grandchildren: Jennifer, Andrea, Joseph and Danielle.

A member of USS *Missouri* Association since 1975, he served as a recording secretary.

Retired from General Electric Co. in 1986 after 39 years of service as quality control inspector in the large steam turbine division.

RICHARD YOUNG, born April 18, 1926 in Trenton, NJ, joined the USN March 9, 1944 and assigned to the USS *Missouri* April 1944, stationed 40mm gun crew and also ran the Adm. Halsey's barge.

While at sea he participated in Leyte Gulf, Iwo Jima, Okinawa and all the Pacific battles they were in.

His memorable experiences include a trip to Rome to see the Pope; the kamikaze plane bringing the delegates to the ship for the signing of the peace treaty. Also seeing the signing of the surrender aboard his ship, the USS *Missouri.*

Young was discharged as S1c May 13, 1946 and was awarded the WWII Victory Medal, American Campaign, Philippine Defense Medal and Asiatic-Pacific w/3 stars.

He has been married to Lillian for 47 years and they have four children: Richard Jr., Linda, Teresa and Thomas and seven grandchildren: Danielle, Shannon, Patricia, Kimberly, Steven and Scott, Robbie. He retired after 40 years in plastics and machine operator and foreman.

PAUL YOUST JR., born Sept. 19, 1923, was inducted into the USN Sept. 19, 1941, Great Lakes, IL. Stations include: Section Base AGC San Francisco, CA; NSDP Base, San Diego, CA; AGC Rec Sta-SO Brooklyn, NY; USS *Asteriou* NYD Washington, DC; USS *Missouri;* Advanced Electric Hydraulic six weeks.

He was awarded the American Theatre Medal, American Defense w/3 Bronze Stars, Asiatic-Pacific and the Good Conduct Medal.

Off duty education courses completed honorable service USN discharge buttons and discharge emblems issued. He held ratings as AS, S2c, S1c, GM3c, GM2c and GM1c (T). Foreign and sea service WWII, honorably discharged from USN, Personnel Separation Center, Great Lakes, IL, Oct. 24, 1945. He was there for the signing of the peace treaty. His wife was on ship topside for the ceremony.

He married Betty Feb. 12, 1944 and they had four children: Darlene, Paul Henry, Rebecca and

Charles. Darlene was born the day the Japanese suicide plane hit side of ship, March 16, 1945. Paul, June 21, 1946; died June 24. Rebecca, Sept. 28, 1947 and Charles, Jan. 12, 1949.

After Youst was discharged he learned the plumbers trade. After four years joined the union in Dayton, OH. Worked that trade until he retired at age 62 and then got sick two years later with cancer. He passed away Aug. 11, 1987. He had six grandchildren and eight great-grandchildren.

MURRAY YUDELOWITZ, born March 30, 1921 in Brooklyn, NY, went into the USN on March 11, 1944. After boot camp at Sampson, NY he was assigned to the USS *Chilton* and then the USS *Missouri.* He was told to be a gunner on a 20mm gun aboard the *Big Mo* was in the 8th Div. His battle station was back aft, on the port side. He was chosen to be the captain's driver. He drove for three captains. He is a plank owner and a plaque owner. He was a S1c when he was on the *Missouri* and was honorably discharged with same rate at Lido Beach, NY on May 13, 1946.

His awards include the Victory Medal, American Theatre Medal and Asiatic-Pacific Medal w/3 Stars.

After the war he went back to his old job in wholesale whiskey and was there for 32 years; 20 years as foreman. He retired in 1989 and is now driving a school bus. He has been married for close to 51 years and they have two children.

JACOB D. ZELDES, born Oct. 10, 1929 in Galesburg, IL, joined the USN in 1947 and was assigned to the USS *Missouri* 1951-1953, 6th Div. Officer, C10 officer. He participated in Korean Campaign and was discharged in 1953 as lieutenant (jg). His awards include the Korean Service, UN Service and South China Sea. His memorable experience includes the sudden death of Capt. Edsel. Zeldes is married and has three children and one grandchild. He is practicing law.

ROBERT V. ZIMBELMAN, born Nov. 12, 1921 in Duncombe, IA, joined the USN March 4, 1941 and was assigned to the USS *Missouri* 1946-1947, as machinist. While at sea he was part of returning the body of the Ambassador of Turkey to his homeland. He was discharged March 3, 1947 and received the WWII Victory Medal, American Defense and American Area.

Zimbelman married Netta Lavender Sept. 30, 1950 and they have two daughters, Linda S. Semler and Ruth A. Espinoza; two grandsons, Robert Espinoza and Brent Semler; two granddaughters, Jaime Williams and Cynthia Espinoza. He retired Jan. 1, 1987 from Garrett-Allied Signal after 30 years as design engineer.

ASSOCIATION MEMBERSHIP ROSTER

This is the roster as provided to the publisher by the Association. The publisher regrets it cannot be responsible for any errors or omissions.

Members and Known Ship's Company

LAST NAME	FIRST NAME
ABAN, JR.,	ARTEMO E.
ABBOTT	ALDEN S.
ABELE	JOSEPH
ADAMS	STANLEY L.
AHL,	BENJAMIN N.
AKERS	MATTHEW L.
ALBANESE	ADOLPH P.
ALBERT,	CLARENCE
ALBERT, SR.,	ARTHUR C.
ALCORN, SR.,	DONALD M.
ALESSANDRO,	TONY
ALEXANDER,	MATTHEW A.
ALEXANDERSON,	ELWOOD E.
ALLEN,	BURTON
ALLEN,	JEROME
ALLEN,	JOHN
ALLEN,	PATRICK
ALLEN,	THOMAS I.
ALLEN,	TODD H.
ALMEIDA,	GEORGE V.
ALONGI,	VITO V.
ALVITI,	WILLIAM H.
ALXANDER,	JAMES F.
ALZAMORA,	ADRIAN F.
AMENDOLA,	JOSEPH P.
AMSTUTZ,	RICHARD G.
ANDERSON,	ADOLPH P.
ANDERSON,	JOHN R.
ANDERSON,	MAYNARD
ANDERSON,	MICHAEL
ANGARANO,	ANTHONY
ANTONIO,	ROBERT J.
ANTOS,	ANATOL
APGAR,	GREGORY
APOSTOLIK	ALBERT A
APPLEGATE	JOHN H.
ARCHER	RICHARD
ARDITO	CHARLES N.
ARENSBERG	JAMES E.
ARGIROS	JAMES G.
ARMSTEAD	MORRIS
ARMSTRONG	CARL E.
ARMSTRONG	EVERETH C.
ARMSTRONG	RAYMOND H.
ARMSTRONG, JR.,	SAMUEL M.
ARONE,	RALPH R.
ARONOWITZ,	ROBERT
ARRINGTON,	TROY A.
ARRUDA,	RICHARD
ATWOOD,	ROBERT E.
AUGUSTINONI,	ARMONDO
AURAS,	ROBERT
AUTINO,	ELIO
AVERITTE,	M.B.
AVERY,	AROLD E.
AVITABLE,	ALEX J.
AYERS,	THOMAS P.
BABAKIAN,	GEORGE
BABBIT,	SAMUEL
BABIARZ,	MARION
BACKER,	DONALD
BADGLEY,	THOMAS W.
BAIN,	GERALD
BAKA,	JOSEPH
BAKAS,	GARY
BAKER,	DEAN F.
BALDUCCI,	DANIEL
BALDWIN,	VINCENT J.

BALFOUR,	ROBERT L.
BALL,	JAMES R.
BAMBILLA,	RENATO G.
BANKHEAD,	REED
BANTA,	NORMAN J.
BARBARO,	PHILIP A.
BARIA,	RAY
BARKER, JR.,	DANIEL S.
BARKER, JR.,	LANDER F.
BARNES,	MAURICE C.
BARNES,	TOBY
BARNWELL,	JAMES M.
BARQUINERO,	PAUL C.
BARR,	JOHNNIE
BARRAZA,	LARRY A.
BARRETT,	VICTOR A.
BARRON,	JOHN C.
BARRY,	RALPH
BARTHOLOMEW,	ALBERT A.
BARTHOLOMEW,	C/O DONNA MORANS
BARTLETT,	DWAYNE N.
BARTON,	MICHAEL
BARTSCH,	VINCENT
BASS, JR,	LIONEL L.
BATISTA,	ORLANDO D.
BAYS,	MRS. MARGARET
BEAL,	THOMAS R.
BEATTIE,	RICHARD
BEATY,	QUENTIN F.
BECK,	DONALD LEROY
BECK,	GERNNIN
BECK,	RAYMOND L.
BECKER,	HARRY E.
BECKETT,	FRED
BEEBE, SR.,	LLOYD
BEEBY, SR.,	LLOYD W.
BEHLING,	JAMES E.
BEIERWALTES,	GUSTAVE A.
BEISTEL,	GEORGE
BELL,	JOHN E.
BELL,	USNR CAPT. JOHN B.
BELLAMY,	LYNN A.
BELLAN, JR.,	JOHN A.
BENDER,	ANTHONY
BENDER,	EUGENE P.
BENNETT,	PAUL B.
BERGER,	WARREN S.
BERGERON,	CURTIS
BERGERON, SR.,	CLIFFORD S.
BERGREN,	USN LCDR RICHARD J.
BESEMER,	ALBERT
BESSERMAN,	MARK
BEST,	FRANCIS M.
BETHEL,	RALPH C.
BIHLER,	DONALD
BILAK,	SHAWN
BILLINGS,	MORT
BILLINGTON,	GORDON L.
BINGENHEIMER,	MATT
BISHOP,	DOUGLAS G.
BISHOP,	ROBERT L.
BISHOP JR.,	D.W.
BISTERSKY,	MICHAEL
BLACK,	DAVID M.
BLACK,	RAY
BLAIR,	JOHN L.
BLAIR,	MIKE
BLAKE,	BENJAMIN
BLAKE,	BERNARD W.
BLANCHARD,	RICHARD A.
BLANCHE,	JOHN N.
BLEDSOE,	HARLEY T.
BLOCH,	DONALD H.
BLONCHE,	JOHN E.

BLUME, JR.,	C.W.
BOCK,	PAUL J.
BOGDAN,	ALFRED
BOGGS,	HERBERT R.
BOLDUC,	ROBERT L.
BOLES,	JAMES E.
BOLLINGER,	JOHN C.
BONERI,	FRED
BONHAM,	IVAN D.
BONILLA,	CARLOS E.
BONNETT, JR.,	DONALD P.
BONUS,	LOUIS
BOOKER,	WILLIAM
BOOKSTAVER,	DR. NELSON D.
BOOTH,	RAY E.
BOOTHROY,	DOYLE E.
BORDEN,	JAMES F.
BORMAN,	WILLIAM G.
BORNKOUSE,	EARL W.
BORREL,	FRANK
BORRELL,	FRANCIS A.
BORRESEN,	C. ROBERT
BOSCHINI, SR.,	VICTOR J.
BOSLET,	CHAPLAIN
BOSTWICK,	GAIL
BOUDREAU,	ED
BOUDREAU,	ROMEO A.
BOWDEN, JR.,	LEWIS B.
BOWENS,	BRIAN EDISON
BOWLES,	JOHN R.
BOWMAN,	LEONARD
BOYLAN,	MICHAEL F.X.
BOYLE,	CMDR.
BOYLE,	DONALD B.
BOYLES,	CLARENCE R.
BRADBURN,	JOHN E.
BRADLEY,	DONALD M.
BRADLEY,	WILLIAM B.
BRADY,	RICHARD B.
BRAGA, SR.,	MANUEL M.
BRANTHOOVER,	W. THEODORE
BRASFIELD,	ROBERT G.
BRAY,	ROBERT C.
BRAZ,	J.J.
BRENNAN,	FRANCIS A.
BRENNAN,	JACK
BRENNAN,	JOHN
BRENNER,	WILLIAM
BRINK,	ROBERT E.
BROADLEY,	PAUL W.
BROCCO,	RICHARD
BROCHHAGEN,	ROBERT J.
BROCK,	JACK
BRODE,	JAMES W.
BRODIE JR.,	CAPT.ROBERT
BROGAN,	F.A.
BROMLEY,	JOHN
BROOKS,	HORACE N.
BROWER,	DOANLD E.
BROWN,	CAPT. WILLIAM D.
BROWN,	DICK
BROWN,	EARL B.
BROWN,	EDWARD
BROWN,	J.P.
BROWN,	JAMES S.
BROWN,	ROBERT A.
BROWN,	RUSSELL D.
BROWN,	SGT. TIMMIE
BROWN,	TOM
BROWN,	WILLIAM J.
BRUBAKER,	CLIFTON M.
BRUNDRIDGE,	DONALD J.
BRUNO,	ROBERT
BRUNS,	DAVID E.

BRYAN,	CHRISTOPHER O.	CHAMBERS,	CARL	CORNWALL,	HARRY T.
BUBNASH,	GEORGE	CHANG,	CHAI TE	CORRADO,	MICHAEL J.
BUCHANAN,	HUGH	CHAPLINE,	EDWIN	CORREA,	JORGE
BUCKLAND,	ROY F.	CHAPMAN,	EDWARD G	CORREA,	MEL
BUCKS,	CHESTER	CHAPMAN,	GEORGE W.	CORSETTI,	LOUIS A.
BUFFMAN,	ED	CHAPUT,	DONALD L.	CORSI,	VINCENT
BUGA,	MARC	CHARPENTIER,	FRED	CORTELYOU,	GEORGE D.
BUIAK JR.,	PETER	CHELF,	RALPH L.	COSTA,	MRS. CHARLOTTE
BULAND,	MAYNARD	CHERNESKY, JR.,	CAPT. JOHN	COUEY,	RALPH
BULLOCK,	RAUL H.	CHERRY,	DON	COUGHLIN,	KENNETH
BUMGARDNER,	JOHN I.	CHIERO, JR.,	BRIAN D.	COULTAS,	RICHARD
BURCH,	DONALD C.	CHOUINARD,	CHARLES C.	COUTROUBIS,	PETER
BURCH,	GEORGE W.	CHOYCE,	NICHOLAS	COUTURE,	ROBERT
BURGER,	KENNETH	CHRISAFIS,	THOMAS S.	COUTURE,	ROLAND O.
BURKE,	DONALD J.	CHRISTIANO,	EDWARD C.	COVERT, JR.,	WILLIAM C.
BURKE,	FRANCIS X.	CHRISTMAN,	WAYNE G.	COVEY,	DONALD W.
BURKE,	MATTHEW F.	CHURGIN,	DONALD	COX,	CLIFFORD C.
BURKE,	MICHAEL E.	CINQUEMANI,	JOSEPH	COX,	STEPHEN
BURKHART,	PETER C.	CIRCELLI,	ALBERT	COX, JR.,	CLYDE
BURN,	JOSEPH T.	CIRIELLO,	ALBERT A.	CRAIG,	STANLEY W.
BURNETT,	MARC	CIRILLI,	JOSEPH P.	CRAIN,	ANTHONY
BURNS,	WILLIAM G.	CLANCY,	MRS. PATRICIA	CRAWFORD,	DARRYL
BURR,	DONALD	CLANTON,	LESLIE B.	CREAMER,	ERVIN D.
BURSCH,	ARNOLD	CLARE,	FRANK	CRISCUOLI,	NICHOLIS
BURT,	SEAN M.	CLARK,	JOHN B.	CRIST,	DONALD
BUTLER,	GEORGE S.	CLARK,	MATTHEW	CRONIN,	HUNTER L.
BUTRYN,	STANLEY J.	CLARK,	MORRIS A.	CROSS,	DON F.
BYINGTON,	COY	CLARK,	RAYMOND	CROSS,	EDWARD
BYRNE,	CHRISTOPHER E.	CLARK,	WILLIAM	CROW,	MILTON F.
BYRNE,	JAMES E.	CLARKE,	FRANK J.	CROWTHER,	DAVID W.
CADORETTE,	RICHARD L.	CLASING,	CARROLL E.	CROZIER,	JEFFREY S.
CAFFEY,	JOCM(SW)JOHN L.	CLAWSON,	FLOYD	CUDDY,	WILLIAM J.
CAIN, JR.,	BURNETT H.	CLAY,	FREDERICK	CUMMINGS,	EDWARD J.
CALDWELL,	ALAN S.	CLAYTON,	RICHARD	CURLEY,	JOHN E.
CALLAGHAN,	ADR WILLIAM	CLEM,	NORMAN	CURLEY,	JOSEPH
CALLAHAN,	JAMES T.	CLEMENS,	ROBERT	CZESLOWSKI,	JOSEPH
CALLAHAN,	JOHN D.	CLEMENT, III,	WILLIS E.	D'ANDREA,	JOSEPH J.
CAMP,	KENNETH A.	CLEMMER,	JEFFERY G.	DAHLER,	FRED
CAMPAGNUOLO,	FRED	CLEVENGER,	GARNET	DALE,	FRANCIS R.
CAMPANARO,	JOSEPH J.	CLIFFORD,	JAMES	DALLAS,	MATHEW
CAMPBELL,	CLARENCE	CLOUSNER,	PHILIP L.	DALTON,	JOHN J.
CAMPBELL,	EDWARD P.	CLOUTIER,	ROBERT M.J.	DAMIENS,	MARCEL J.
CAMPBELL,	FORREST	CLUCK,	JAMES E.	DAMM,	D.C.
CAMPBELL,	LEE	COFFEY,	ROBERT F.	DANIEL,	CWO4 DWIGHT
CAMPBELL,	ROBERT S.	COFFEY,	WILLIAM T.	DANIEL,	FRED G.
CANBY,	FREDERICK L.	COLBURN,	HERBERT T.	DANIEL,	JOHN R.
CANCILLIERE,	FRANK	COLE,	JEFFREY D.	DARRAH,	RANDY
CANNELLA,	VINCENT J.	COLE,	PHILLIP	DASHER,	RALPH C.
CANNEY,	MALCOLM H.	COLE, III,	FRED W.	DAVID,	HARRY S.
CANNOP,	RALPH	COLLINS,	ANDREW T.	DAVID,	JOSSIE T.
CAPORASO,	NICHOLAS J.	COLLINS,	CHAD	DAVIDSON,	JOHN M.
CARBER USN,	LCDR F.H.	COLLURA,	VINCE	DAVIS,	E.G.
CARDEN,	GEORGE	COLTON,	ARTHUR E.	DAVIS,	EUGENE A.
CARENZA,	JOHN L.	COLWELL,	CMDR. JOHN B.	DAVIS,	EVERETT L.
CAREY,	RALPH W.	COMPETIELLO,	TEPHEN	DAVIS,	WILLIAM
CARICO,	JOHN	CONCILLA,	DOMINICK	DAVIS JR.,	DONALD D.
CARIGNAN, JR.,	RICHARD C.	CONLEY,	FRANCIS J.	DAVITT,	JOHN W.
CARLSON,	CHARLES E.	CONNER,	RICHARD E.	DAWSON,	GEORGE
CARLSON,	FRANK W.	CONNLY,	JOHN J.	DAY,	STEVEN
CARLSON,	LT. JOHN	CONNOR,	DONALD R.	DAYBERRY,	JOHN E.
CARMINT JR.,	ROBERT W.	CONRAD, JR.,	ANDREW H.	DE ANGELIS,	LOU
CARMON,	KEVIN	COOK,	CHARLES M.	DE FAZIO,	DAMON
CARNEY,	CAPT. JAMES	COOK,	DANIEL	DE FILIPPIS,	TONY
CAROLIN,	MICHAEL H.	COOK,	PAUL D.	DE LEO,	JOHN
CARON,	RICHARD	COOK,	ROBERT L.	DE PARIS,	MICHAEL
CARRIGAN,	EDWARD J.	COOKE,	HARRY F.	DE SANTIS,	LCDR ROBERT W.
CARTER,	JOSEPH L.	COOKE, JR.,	SIDNEY	DE SAULNIERS,	LAWRENCE B.
CARUSO,	CALOGER J.	COOLEY,	NORMAN V.	DE SOUZA, JR.,	MANUEL R.
CASEY,	BERNIS J.	COOLIDGE,	JERRY	DE VINNEY, JR.,	LLOYD J.
CASEY,	DUANE F.	COON,	JAMES W.	DEELEY,	CHARLES C.
CASEY,	RICHARD L.	COON,	KENNETH L.	DEGLIO,	NICHOLAS
CASSIDY,	JOSEPH	COON, JR.,	E. RALPH	DEITZ,	WALTER F.
CASSON JR.,	WALTER A.	COONEY,	CHRISTOPHER A.	DELANEY,	JOHN J.
CAVANAUGH,	FRANCIS	COONTZ,	ROBERT J.	DELLARIPA,	LOU
CEGELIS,	MATTHEW	COOPER,	ERVIN H.	DEMARCO,	PHILIP J.
CEGELKA,	JOSEPH	COOPER,	HARVEY E.	DEMERS,	MARK E.
CELANO,	USN CDR JOSEPH	COOPER, SR.,	GEORGE	DEMOS,	PAUL G.
CELLI,	PASQUALE A.	COPPLE,	CLYDE N.	DEMPSEY,	GEORGE F.
CEMBALSKI,	ALEX B.	CORK,	ORVILLE L.	DENEAU,	LYLE P.
CHALFIE,	ELI	CORLEY,	LT. DAVID D.	DENNISON,	CAPT. ROBERT L.

DENTON,	WILLIAM	EDMONDS,	ROBERT A.	FLIGHT,	STANLEY
DEPHILIPPO, SR.,	EDWARD O.	EDSALL,	CAPT. WARNER R.	FLORES,	REYANLDO
DERESKY,	JOSEPH L.	EDWARDS,	TOM	FLORI,	ADAM
DESAULNIERS,	PAUL C.	EFFINGER,	FRED	FLORY,	JOSEPH C.
DESCAMPS,	CYRIL M.	EGAN,	PETER C.	FLYNN, JR.,	ROBERT W.
DEVORE,	JOHN C.	EGGAN,	MARTIN	FOLEY,	WILLIAM J.
DEWAR,	ARTHUR A.	EICHENLAUB,	BOB	FORD,	JACK M.
DEXTER,	GEORGE	EICHMAN,	GEORGE	FORD,	JOHN B.
DEXTER,	IVAN	EIFRID,	JOSEPH O.	FORNITO,	CARMEN
DI MATTEO,	VINCENT A.	ELCESS,	DAVID C.	FORSETH,	ROBERT
DI SANTO, SR.,	MICHAEL A.	ELLIOTT,	WYNE E.	FORTIER,	LUCIEN J.
DI SAVINO,	ROBERT J.	ELLIOTT, JR.,	JAMES W.	FORTIN,	ROBERT J.
DIAL,	JOHN	ELLIOTT, JR.,	ROBERT W.	FORTNEY,	FRED H.
DICKERSON,	GUY	ELLITHORPE,	GORDON (BUCK)	FOSBURG,	DONALD R.
DICKINSON,	GORDON L.	ELMER,	MAX L.	FOSSER,	PHILIP J.
DICKMAN,	ROBERT M.	EMORY,	WILLIAM C.	FOSTER,	DAVID B.
DIETRICH,	ROBERT	ENGELBRECHT,	DALE W.	FOURNIER,	PAUL
DIFFENPAL,	ROBERT	ENGLER,	ROBERT	FOX,	LOUIS L.
DILIDDO,	ROBERT F.	ENGLISH,	GERALD	FOX,	MARSHALL E.
DILL,	ROLAND H.	ENGVALSON,	MAJ. PAUL W.	FOX,	NORMAN
DIMAIO,	VITO N.	ERDMAN,	FLOYD G.	FRAILEY,	GORDON L.
DIMOCK,	DANIEL E.	ERPELDING,	JOHN W.	FRAME,	RYAN J.
DIONNE,	ANGENAR	ERTLMEIER,	GEORGE	FRANKLIN,	DONALD R.
DITTO,	WILLIAM L.	ERWIN,	ERNST	FRANZ,	RICHARD
DIXON,	ARTHUR G.	ESCALON, JR.,	VICTOR	FRASER,	CHARLES E.
DIXON,	WILLIAM A.	ESCHLIMAN,	RONALD D.	FRASER,	STEWART A.
DOBSON,	FRANK A.	ESDEN,	JAMES A.	FRAZIER,	WILLIAM R.
DODD,	RAY J.	ESTRADO,	ANTONIO	FREDRICK,	CHARLES
DOERING,	ALFRED S.	ETSCHEID,	RHEIN JAMES	FREDRICKS,	DONALD
DOHLER,	ARTHUR A.	ETUE,	MRS. PEARL	FREDRIKSON,	SCHUYLER
DOLAN,	LEONARD R.	EVANS,	ALMON G.	FREEMAN,	JOSEPH G.
DOLCE,	SAM	EVANS,	DAVID W.	FREEMAN, JR.,	CLARENCE C.
DONATELLI,	EDWARD	EVANS,	GORDON B.	FREESE,	WILLIAM D.
DONLEY, JR.,	HARLEY	EVANS,	ROBERT K.S.	FREZZA,	ANTHONY
DONOVAN,	EDWARD C.	EVANS,	WILLIAM T.	FRIEDLAND,	AARON J.
DOOLEY,	J.M.	EVERHEART,	DONALD R.	FRIEDRICH,	DOUGLAS A.
DOOLEY,	JOHN W.	EWBANK,	CARLTON L.	FRIELAND,	AARON
DOONG,	CDR. LAWRENCE K.C.	EWENS,	ROBERT S.	FRITSCH,	WILLIAM C.
DORAN,	JAMES	EWING,	EDWARD W.	FROST,	JAMES O.
DOREY,	ERNEST L.	EYER,	GEORGE B.	FROST,	THOMAS
DORKO,	ROBERT S.	FAHR, JR.,	HERBERT	FROTHINGHAM,	LARRY
DOWELL,	T.M.(TED)	FAHY,	ANDREW W.	FRUCHT,	CARL F.
DOWLEY,	EDWIN R.	FAIR,	DEWAYNE	FULKERTH,	DANIEL J.
DOWNEY,	JOSEPH C.	FAIRBANKS,	RAYMOND	FULLERTON,	C.L.
DOYLE,	MRS. ANN S.	FAIRFIELD,	DOUGLAS	FULTON,	DAVID A.
DRAPER,	ALMER	FAJNOR,	JOHN P.	FUNDERBURKE,	WILLIAM H.
DREIER,	LT USN ARMIN F.	FALKENBACH,	GILBERT	FUOCO,	FRANK
DRIVER,	JAY	FALVO,	FRANK	GABEL,	BRUCE R.
DU BOIS,	FREDERICK J.	FARIS,	EUGENE L.	GABRYSZAK,	ALEX
DUBEE,	MRS. GAIL	FARMER,	LEROY	GAGLIARDI,	DOMINICK
DUBENSKY,	JOHN A.	FARNSWORTH,	ROBERT L.	GALGANO,	VINCENT
DUBOIS,	JOSEPH H.	FARR,	BERT	GALLAGER,	ED
DUBSKY,	JOSEPH A.	FAULK,	CHAPLAIN	GALLEGOS,	TERRY
DUCK,	DON W.	FEIGENBAUM,	JACK	GALLI,	LOUIS
DUCKETT,	DONALD R.	FELLOWS,	WILLIAM	GALLO,	PASQUALE
DUDLEY,	LT. JG S. MARK	FELS,	IVAN	GALVIN,	ROBERT E.
DUFFY,	DOUGLAS B.	FENTON,	FRANCIS I.	GAMBLE,	GEORGE
DUFFY,	JOHN R.	FERGURSON,	WALTER S.	GAMBLE,	RADFORD "BUD"
DUGAN,	LAWRENCE J.	FERGUSON,	MATTHEW J.	GAMBLE,	STEPHEN K.
DUKE,	CAPT. IRVING T.	FERRAINA,	DOM ATTY	GAMBOE,	JAMES H.
DULA,	NOAH J.N.	FERRANTE,	KENNETH T.	GANAS,	JULIAN J.
DUMAS, JR.,	ROBERT	FERRER,	EDDIE M.	GARBER,	FRED
DUNCAN,	ROBERT E.	FERRIGNO,	PASQUALE	GARCIA,	FRANK
DUNHAM,	THEODORE H.	FIERS,	ROBERT H.	GARCIA,	TIMOTHY
DUNLAP,	WILLARD F.	FILLORAMO,	FRANK	GARDNER,	ANDRE K.
DUNN,	HENRY W.	FINCHER,	JEFFERSON C.	GARNER,	MALONE A.
DUNN,	JOSEPH L.	FINE,	HARRY O.	GARRINO,	EPARAIMS
DUNNE,	JAMES M.	FINLEY,	EUGENE G.	GARZCRELLI,	JOHN
DUNNING,	NORMAN R.	FINN,	MICHAEL P.	GASKELL,	FREDRICK J.
DURFEE,	BRETT	FIORILLO,	JOHN L.	GATO,	JOSEPH S.
DURKIN,	JOHN	FITZGERALD,	JAMES G.	GAUDET,	JOSEPH J.
DUX,	FRANK J.	FITZGERALD,	LCDR SUSAN	GAULIN,	NORMAN E.
DYE,	STEVE	FLAHERTY,	DONALD	GAY,	JOHN M.
DYER,	JEFFERY A.	FLAMM,	DR. PAUL A.	GEAMPAOLA,	ANGELIO
DYER, SR.,	F. WILLIAM	FLANAGAN,	NICHOLAS E.	GEARS,	CARL E.
EATON,	ERNEST E.	FLAWD, JR.,	HARRY E.	GEISSLER,	GILBERT B.
EATON,	JACK	FLEENER,	BUDDY R.	GEMELLI,	JOSEPH J.
EBBERT,	YN3 ROBERT	FLEMMING,	ROBERT A.	GENDREAU,	RAYMOND J.
EBBS,	KAREY	FLETCHER,	EDDIE	GENEREUX,	WILLIAM
EDDY,	HOWARD R.	FLETCHER,	PAUL L.	GENEST,	ROBERT A.

GENOVESE,	SAMUEL	HACKWORTH,	DAVID	HIGHTOWER,	HARRY
GEORGE,	CLARENCE M.	HADLEY,	PHILIP	HILDERBRAND,	MRS. KENNETH
GERBER,	ALVIN	HAHN,	CDR. RICHARD C.	HILL,	CAPT. TOM B.
GERDES,	GERALD E.	HAKEL,	LAWRENCE	HILL,	FRANCIS
GEREN,	PATRICK	HALL,	MICHAEL S.	HILL,	PETER A.
GERHARDT,	ROBERT	HALL,	ROBERT	HILL,	THOMAS C.
GERKE,	WILLIAM EDWARD	HALLER,	DONALD T.	HILLENKOETTER,	V ADM ROSCOE H.
GERSHMAN,	BERNARD G.	HAMMEL,	RALPH	HILLIARD,	JIMMIE
GERTLER,	RAY	HAMMEL,	WILLIAM R.	HIMSEL,	DENNIS M.
GIBLIN,	THOMAS J.	HAMMOND,	FRANK W.	HINNERS,	LOWELL
GIBSON,	CHARLES E.	HAMMOND,	JAMES E.	HINSON,	BRODUS W.
GIBSON,	L.D.	HAMMONDS,	CHARLES L.	HINSON,	HARVEY C.
GILES,	PHILIP	HAMPTON,	JAMES C.	HO,	HUY A.
GILLESPIE,	PERRY C.	HANCOCK,	WESLEY A.	HOBART,	ROBERT
GILLIGAN,	CHARLES	HAND,	MARVIN R.	HOBERG,	HENRY C.
GILLIS,	DONALD	HANDS,	MRS. EDWARD	HOBIN,	WALTER E.
GILLISPIE,	BILLY RAY	HANEY,	CHARLES J.	HODGE,	ELDRED
GILMAN,	KENNETH H.	HANNIGAN,	CORNELIUS	HOFMAN,	PNCM T.C.
GIUSTI,	VINCENZO	HANSEN,	CHARLES	HOGAN,	LARRY
GIVENS,	GOLDIE	HANSEN,	WILLIAM L.	HOKE,	JEREMY R.
GLAUBITZ,	GERALD A.	HANSEN, JR.,	ELINAR	HOLCOMB,	WILLIAM L.
GLEASON,	JOSEPH K.	HAPENNY,	JAMES M.	HOLLAND,	FRANCIS P.
GLOGER,	LAVERN	HARBOURNE,	WILLIAM H.	HOLLAND,	RICHARD
GLOGG,	GEORGE J.	HARDER,	RICHARD C.	HOLMES,	LOWELL G.
GLYNN,	THOMAS P.	HARDY,	DAVE	HOLTON,	CHARLES S.
GODFREY,	EDWARD W.	HARFF,	FRANCIS J.	HOOVER,	DARRYL
GODWIN,	GLENN	HARKNESS,	RICHARD R.	HORAN,	WILLIAM F.
GOFFREDO,	ANGELO S.	HARRIMAN,	GEORGE	HORN,	BERTRAM
GOLDEN,	HAROLD S.	HARRINGTON,	BILLY M.	HORN,	JACK
GOLDMAN,	FREDDIE	HARRINGTON,	JOHN P.	HORNSTEIN,	STANLEY
GOLLY,	MARK A.	HARRINGTON,	VERNE P.	HOROWITZ,	ELLIOT
GOLSON,	BILLY J.	HARRINGTON,	WILLIAM L.	HORTON,	GEORGE H.
GONYA,	LAWRENCE	HARRIS,	A. RICHARD	HOSKINS,	CHARLES E.
GOOD,	KENNETH	HARRIS,	LES	HOSTEIN,	MILTON
GOODMAN,	LAWRENCE E.	HARRIS,	RICHARD E.	HOUGHTON, III,	AMORY M.
GORDON,	JOHN	HARRISON,	GEORGE W.	HOUSE,	MM1 ERNEST L.
GORDON,	ROBERT M.	HARRISON, JR.,	LT. WILLIAM T.	HOVORKA,	LAWRENCE D.
GORHAM,	BILL	HART,	EDWARD A.	HOWARD,	SANFORD A.
GORMLEY,	JAMES	HARTMAN,	ROY JAMES	HOWDESHELL,	TONY L.
GORMLEY,	JAMES P.	HARTNETT,	DANIEL J.	HOWELL,	GEORGE H.
GOSNELL,	JAMES B.	HARTSON,	ALMON CECIL	HUBBARD,	GEORGE F.
GRAFF,	ELWOOD H.	HARVEY,	ROBERT R.	HUBER,	NELSON A.
GRAGG,	STEVE	HARVEY, SR.,	EDWIN D.	HUFFMAN,	CHARLES
GRAHAM,	DAVID C.	HASSETT,	BERNARD A.	HUGHES,	JAMIE WALTER
GRAHAM,	GENE G.	HATCH,	PAUL O.	HUGHES,	RICARDO F.
GRAHAM,	ROBERT D.	HATCH,	ROBERT W.	HULA,	RAYMOND W.
GRAVES,	CALVEN D.	HATFIELD, II,	DEAN P.	HULEN,	GLENN
GRAY,	JAMES EDWARD	HATTON,	LESLIE O.	HULL,	JOHN
GRAY,	ROBERT N.	HAUSNER,	JIM	HUMPHRIES,	AUGUSTAS
GREATHEAD,	LEONARD W.	HAWK,	DON	HUNERKOCH,	MARSHALL S.
GREAVES,	FREERICK W.	HAWXHURST,	BILL	HUNT,	JIM
GREEN,	TARIO M.	HAYDEN,	KEN	HUNT JR.,	GEORGE L.
GREEN,	WILLIAM F.	HAYS,	VERLYN	HUNTER,	D.L.
GREENE,	HOLLIS	HEATH,	LEE	HURST,	JIM
GREENE,	REGENALD O.	HECKERT,	GERALD E.	HURST,	LYLE J.
GREER,	GEORGE L.	HEFTY,	ROBERT T.	HUSSEY,	ROBERT J.
GREGORY,	MICHAEL	HEIDER,	KENNETH J.	HUYSERS, JR.,	CARL
GRENHAM,	USNR REV. JOHN P.	HEIN,	DERRICK R.	HYNES,	JACK
GRIAK,	FRANCIS	HELBIG,	WALTER L.	HYSTAD,	CLIFFORD
GRIFFITH,	RICHARD R.	HELMLY,	ROBERT L.	IACOVONE,	ANGELO
GRIFONI,	VINCENT LEO	HELMSTETTER,	ROBERT F.	INGATO,	VINCENT
GRIMM,	LAWRENCE H.	HEMPHILL,	JARRELL D.	INGLESE,	VINCENT JAMES
GRISSOM,	CHARLES K.	HENDERSON,	RICHARD A.	IRWIN,	GEORGE S.
GROVE,	ROBERT R.	HENKING,	ALFRED M.	ISAACSON,	WILLIAM D.
GRUENEVELD,	WILLIAM	HENRICKS,	EUGENE H.	IVERSON,	KEN J.
GRZYBOWSKI,	FRANK	HENRIQUES,	FRED	JACKSON,	GEORGE P.
GSTOETTNER,	JAMES J.	HENRY,	HARRY J.	JACKSON,	ROBERT N.
GUDDE,	CARY DEAN	HENRY,	IRA A.	JACKSON,	RODNEY W.
GUERTIN,	JOSEPH R.	HENRY,	KEVIN A.	JACOBS,	IAN FRANK
GUINEY,	BILL	HERBERT,	HAROLD G.	JACOBS,	ROBERT J.
GULIANO,	DOMINICK	HERGENRATHER,	ROBERT N.	JACOBSON,	LAWRENCE E.
GUNST JR.,	CYRIL C.	HERGET,	CARROLL A.	JACOBY,	RONALD G.
GURLEY,	MICHAEL	HERRINGTON,	J.H. "TINY"	JAECH,	JUSTIN A.
GURTNER,	ALBERT J.	HERRIT,	KENNETH E.	JAIMES,	JAMMIE A.
GUSTAFSON,	BOB	HERSHBERGER,	DONALD	JAMES,	RUSSELL F.
GUSTAFSON,	RICHARD E.	HERTZ,	JACOB D.	JANKO,	MARVIN
GUSTAFSON,	WALLACE F.	HESS,	ROBERT TRUMAN	JANSEN,	THOMAS J.
GUTHRIE,	JAMES E.	HESSELS,	WARREN	JARMAN,	LLOYD H.
GWALTNEY,	JOHN H.	HEWETT, JR.,	CHARLES A.	JARVIS,	RONALD
HAAS,	BRIAN SCOTT	HICKEY,	HUGH J.	JENKINS,	DONALD E.

JENSEN,	VANCE W.	KNOX,	WILLIAM S.	LINDSAY,	PAUL
JOHNSON,	CHARLES A.	KOERNKE,	ALBERT OTTO	LINDSEY,	EDGAR W.
JOHNSON,	CLARENCE E.	KOGAN,	DENNIS	LINDSTROM,	GERALD A.
JOHNSON,	DONALD C.	KOHLER,	CHARLES	LINDSTROM,	L.W.
JOHNSON,	KEN	KOLENUT,	CHARLES W.	LINETTE, JR.,	ANDREW M.
JOHNSON,	RAY G.	KOLLMAN,	FC2 MARK E.	LINEY,	WILLIAM J.
JOHNSON,	ROY	KOLOTA,	LAWRENCE S.	LIPPY,	CHARLES W.
JOHNSON,	RUSSELL L.	KONOSKY,	ROBERT J.	LITFIN,	WERNER A.
JOHNSON,	TROY D.	KOPP,	KENNETH	LOBA,	ROBERT T.
JOHNSON,	WILLIAM R.	KORN,	EDWARD	LOGUE,MD,	MICHAEL J.
JOHNSON, JR.,	AMOS R.	KORN,	ROBERT	LOISELLE,	RENE F.
JOHNSTON,	RICHARD H.	KORNEY,	USN CDR. JIM	LOMBARD,	ANTHONY J.
JONES,	B.GORDON	KRETSCHMANN,	WALTER	LOPEZ,	ROY J.
JONES,	DOUGLAS M.	KRULISH,	FRANCIS C.	LOPEZ,	TONY F.
JONES,	HENRY	KUHN,	JOHN G.	LORD,	GERALD W.
JONES,	JOHN	KURDELSKI,	GARY A.	LORITZ,	HAROLD
JONES,	JOHNNY	KUVET,	JOHN R.	LORSON,	JOSEPH R.
JONES,	LEWIS C.	LABAMPA,	JESUS A.	LORUSSO,	MICHAEL
JONES,	PETER	LABARRE,	FLOYD D.	LOTHER,	CLARENCE
JONES,	VERNON L.	LACK,	SAMUEL	LOUCH,	DEREK M.
JORDAN,	CAPT. KENNETH S.	LACOSTE,	BRUCE J.	LOUCKS,	WILLIS H.
JOSEPH,	THOMAS	LADNER,	STANFORD L.	LOUDERMILD,	WILLIAM
KACZMACEK,	ZYMONT	LAFAYETTE,	REGINALD P.	LOURENCO,	MANUEL
KAHLE,	JAY	LAFORTY,	GMC CHUCK	LOVE,	RAY
KAIGLE,	LEOPOLD R.	LAGERGREN,	WILLIAM	LOVELAND,	NOEL R.
KAISS,	CAPT. LEE	LAJARA,	JOHN R.	LOW,	JIM
KALANTA,	EDWARD M.	LAKOSKI,	ROLAND A.	LOWERY,	HARRY
KAMET,	TOM	LANCIERS,	PETER A.	LOY, SR.,	MAYNARD E.
KARCHER,	RAYMOND N.	LAND,	EUGENE W.	LUCAS,	DAVID W.
KARCHER,	ROBERT	LANDIS,	BRADLEY A.	LUCAS,	JOHN E.
KARGENIAN,	LEON M.	LANDRY,	LONNIE	LUCAS,	LONNIE
KARO,	GEORGE A.	LANE,	FREDRICK W.	LUND,	FREDERICK W.
KASS,	BERNARD	LANE,	JOHN	LUNTKOWSKI,	FRANCISCO
KATHERMAN,	EARL	LANE,	ROBERT E.	LUX,	DAVID
KATZ,	SAUL	LANGLEY,	R.L.	LYMAN,	GARY
KECK,	ALVIN D.	LANGLEY,	ROGER	LYNCH,	JOSEPH F.
KEENE,	ELMER W.	LANGLOIS,	EM1 JAMES B.	LYNN,	COREY D.
KEITH,	CAPT. ROBERT	LANOYE,	PHILIP J.	MAAG,	EDWARD J.
KEITH,	JEFFREY	LANSING,	GLENN V.	MAC COLL, III,	REV. JAMES R.
KEITH,	PAUL V.	LANTZ,	WILLIAM F.	MAC LAFFERTY,	CONAN E.
KELL,	ERNIE	LASSEN,	WALTER J.	MAC MILLAN,	EDWARD R.
KELLER,	CLEO E.	LATHROP,	ROBERT D.	MACDONALD,	REID B.
KELLER,	HERBERT A.	LATLIP,	LEO A.	MACKEY,	ROBERT G.
KELLER,	VERNON H.	LAVERY,	JAMES P.	MACRIE,	NELSON D.
KELLEY,	ALFRED E.	LAW,	DAVID ARTLEY	MADDEN,	JOHN P.
KELLEY,	EDWARD S.	LAWTON,	EDGAR L.	MADDEN,	MORGAN N.
KELLEY,	EUGENE J.	LAWTON,	MICHAEL C.	MAGEE,	PHILLIP S.
KELLY,	ALBERT J.	LAYNE,	BRUCE F.	MAGILL,	ROBERT M.
KELLY,	CHARLES W.	LE BLANC,	LEO J.	MAHONEY,	CHARLES
KENNEDY,	JAMES T.	LE FORT,	HAROLD E.	MAHONEY,	JAMES
KENNEDY,	ROBERT C.	LEADBEATER,	RICHARD	MAILETT,	E.J.
KENNEMORE,	RUFUS H.	LEAVITT,	EDWARD J.	MAILLET,	ALPHEE
KENNEY,	RICHARD T.	LEAVITT,	SHELDON J.	MAIN,	ROBERT V.
KESSELRING,	WILLIAM T.	LEBLANC,	ALFRED H.	MALEY,	JOHN C.(JACK)
KIGER,	GEORGE V.	LEBLANC,	MRS. KATHIE	MALLISON,	PERCY W.
KILGORE,	KEVIN R.	LECHNAR,	FRANK P.	MALONE,	L.T.
KILLEBREW,	BRUCE	LEE,	CURTIS	MALONEY,	PHILIP T.
KIMMEL, JR.,	CHARLES M.	LEE,	IRA B.	MALPHRUS,	LAVAL
KINDRED,	LAWRENCE	LEE,	WARREN	MANCINI,	DR. FRANK D.
KING,	CHARLIE B.	LEFKOWITZ,	HARRY	MANCINI,	JOHN A.
KING,	ERNEST C.	LEIBIG,	HERMAN B.	MANDEL,	MICHAEL
KING,	GEORGE C.	LEIST,	CLAYTON W.	MANGELS,	WILLIAM
KING,	JOHN W.	LEITZ,	WILLIAM	MANN,	GEORGE
KING,	WALTER C.	LENDACKY,	LARRY	MANN,	ROY S
KING, JR.,	WILLIAM	LEONARD, JR.,	JOSEPH A.	MANNING,	JOSEPH P.
KINGDON,	GEORGE R.	LEONETTI,	JOSEPH	MANNING,	RICHARD S.
KINGHORN,	ROBERT L.	LESLIE,	EDWIN B.	MANZANO,	MAJOR E.B.
KIRBY,	MRS. JOYCE	LESTER,	KENNETH R.	MARADAKIS,	ALEX
KIRKHOFF,	LT.	LESTER,	RUSSELL H.	MARCHESE,	HENRY C.
KIRKLAND,	KEITH W.	LETHGO,	WILLIAM B.	MARKS,	ROBERT C.
KIRKMAN,	JACK	LEWIS,	JOHN	MARLOW,	VIC H.
KISHBAUGH,	BELVIN	LEWIS,		MAROLLA,	PHILIP
KISKINIS,	GEORGE J.	LEWIS,	LIONEL B.	MARSHALL,	HENRY
KITCHINGS,	W.C.	LEZAK,	FRED	MARTIN,	ALEXANDER
KITTOE,	KENNETH K.	LHOTAN,	THOMAS J.	MARTIN,	CURTIS W.
KLOSS,	HARRY J.	LIGUORI,	JAMES G.	MARTIN,	JACK
KLUG,	RICHARD C.	LILLIBRIDGE,	JOHN	MARTIN,	PAUL
KNIGHT,	LARRY L.	LIM,	KIN H.	MARTIN,	R.B.
KNIGHT, II,	HENRY D.	LINDEL,	ROBERT	MARTIN,	TIMOTHY M.
KNOBLE,	RICHARD N.	LINDEMAN,	DONALD J.	MARTINA,	JOSEPH A.

MARTINEZ,	PETER	METZGER,	MATTHEW	MURRAY,	ADM STUART S.
MARTINO,	MICHAEL A.	MEYER,	ROY O.	MURRAY,	ALAN R.
MARTINO,	VIRGIL J.	MEYERS,	ROBERT A.	MURRAY, JR.,	RADM JAMES
MASLIN,	CHARLES W.	MICALE,	ANGELO J.	MYERS,	EDWARD J.
MASON,	WALTER	MICHAUD,	JOSEPH	NAPOLITANO,	FRED
MASON,	WILLIAM J.	MIDGET,	JIMMY	NARCISSE,	WAYNE
MASONE,	JINO	MIDKIFF,	VERNON E.	NAU,	ARTHUR
MATHER,	CHARLES H.	MIELE, SR.,	JOSEPH A.	NELSON,	MARTIN P.
MATHEWS,	ERNEST T.	MIERZEJEWSKI,	ANTHONY	NELSON,	PAUL H.
MATTHAEI,	CHARLES W.H.	MIKESELL,	DAVID R.	NELSON,	RANDY R.
MATTHESEN,	G.S.	MILLER,	AVON M.	NELSON,	ROBERT
MATTHEWS,	MARK S.	MILLER,	DALE KENNETH	NELSON	
MATTRELLA,	JACK	MILLER,	DAVID	NEWHOUSE,	NORMAN
MATZKE,	IRVIN L.	MILLER,	JOHN A.	NEWTON,	JAMES E.
MAXFIELD,	ALLEN F.	MILLER,	LEWIS M.	NEWTON,	KENNETH A.
MAY,	PAUL C.	MILLER,	LYLE C.	NGUYEN,	THO VAN
MAYEN,	CARLOS	MILLER,	TODD	NICHOLS,	ROBERT B.
MAYETTE,	MRS. LOIS	MILLICK,	MICHAEL	NICKOLS,	USN, CAPT JAMES P.
MAYNARD,	JAMES	MILLS-PRICE,	EDGAR J.	NIEHOFF,	CHARLES J.
MAYNARD,	WILLIAM	MINARD,	PHILLIP RAY	NIELSEN,	RALPH J.
MAYNARD, JR.,	M.C.	MINIEA, JR.,	CHIEF SYLVESTER	NOBIS,	ROBERT S.
MC ALISTER,	JACK E.	MINNETTE,	JOHN G.	NOBLE,	PAT
MC CANN,	PAUL	MINOR,	JAMES W.	NOCENTE,	FRED J.
MC CARRON,	JOHN H.	MIRIELLO,	BENJAMIN F.	NOONAN,	WILLIAM J.
MC CARTHY,	ROBERT	MITCHLER,	ROBERT W.	NORBERG,	WILLIAM H.
MC CLELLAN,	JAMES H.	MIXON,	JOHN D.	NORCROSS, JR.,	LCDR MURRAY C.
MC CLELLAND,	CHARLES W.	MLINCSEK,	JOHN	NORRIS,	ROBERT
MC CLOSKEY,	ERNEST R.	MOBILIA,	ROSS F.	NORTH,	CMDR. JAMES R.
MC CLOUD,	JOSEPH T.	MODICA,	ANTHONY	NORTHUP,	GARDINER L.
MC CLURE,	KENNETH A.	MOGAVERO,	ALFRED G.	NOTTINGHAM,	WILBUR I.
MC CLURE,	ROBERT	MOISE,	EUGENE	O'BRIEN,	JOHN
MC COLLUM,	BERNARD N.	MOITY,	HENRY C.	O'BRIEN,	JOHN W.
MC CONNELL,	JAMES J.	MOLOPALSKI,	THEADORE	O'MALLEY,	MARTIN
MC COOL,	WHITTIE J.	MONEY,	JAMES W.	O'NEIL,	ROBERT
MC COY,	CHARLES M.	MONTGOMERY,	CHARLES	O'REILLY,	ROBERT W.
MC DONALD,	CORNELIUS P.	MOODY,	JEREMY D.	O'TOOLE,	LEO J.
MC FADDEN,	ROGER W.	MOORE,	ARTHUR P.	OBERTS,	JAMES E.
MC FARLAND,	DAVID D.	MOORE,	DONALD	OBITZ,	WILLIAM
MC GEE,	EDWARD J.	MOORE,	MORRIS	OCKER,	CALVIN
MC GHEE,	WILLIAM	MOORE,	PAUL A.	OCORR,	DAVID
MC GINNIS,	JAMES P.	MOORHEAD,	C.A.	ODELL,	JAMES
MC GOUGH,	PATRICK J.	MOOSE,	FCCS(SW)ROBERT	ODOM,	EARL B.
MC GRATH,	VINCENT P.	MORAN,	LOUIS G.	OJA,	GLENN N.
MC GUINNESS,	JAMES A.	MORANO,	JOHN R.	OLIVA,	FELIX S.
MC HUGH JR.,	JOHN J.	MORENO,	JOSE R.	OLIVER,	DAVID C.
MC INTOSH,	MILTON L.	MORGAN,	ALFRED C.	OLSON,	HARLAN N.
MC KELL,	DAVID F.	MORGAN,	DOC	OLSZEWSKI,	EDWARD
MC KIERNAN,	RALPH	MORGAN,	RANDALL	ONDRAKO SR.,	THOMAS J.
MC KINNEY,	JOHN D.	MORIN,	LEO H.	ONISCHUK,	WALTER A.
MC LELLAN,	A. STUART	MORIN,	WILLIAM	ORBAN,	FRANK P.
MC LELLAN,	MACK H.	MORLAN,	PATRICK L.	ORBAND,	DOMINICK
MC LOUTH,	DONALD	MOROSS,	GEORGE M.	ORLANDO,	JOSEPH
MC MANUS,	PAUL J.	MORRE,	ROBERT E.	OSENBACH,	THOMAS
MC VAY,	LOUIS R.	MORRELL,	STEPHAN G.	OSTEEN,	HASKELL
MCCARTHY, SC,	USN, RADM. J.D.	MORRIS,	DAVID WAYNE	OSTERHOUDT,	IRVING C.
MCCLURE,	KENNETH A.	MORRIS,	RALPH G.	OTTEN,	JAMES
MCCORMICK,	JOHN D.	MORRIS, JR.,	THOMAS W.	OTTMAN,	LYMAN J.
MCCRINDLE,	ROBERT D.	MORRISON,	CLYDE E.	OVERTON,	VERNON M.
MCDONALD,	HARRY D.	MORRISON,	DONALD A.	OWENS,	ALLEN S.
MCKINNEY,	JOHN D.	MORRISON,	JOHN C.	OWENS,	JAMES R.
MEAD,	HAROLD R.	MORSE,	EUGENE H.	PACHELO,	ROBERT E.
MEADOWS, JR.,	BOBBY G.	MORSE,	LOUIS B.	PADESKY,	THOMAS
MEALEY,	JOHN J.	MORSE,	RAY	PAKER, JR.,	JIMMY
MEDINA,	MICHAEL J.	MORTON,	WILLIAM D.	PALATUCCI,	ANGELO M.
MEEHAN,	PATRICK S.	MOSELEY,	RAYMOND H.	PALDEZ,	JESSE E.
MEENAHAN,	JOHN P.	MOSER,	DONALD R	PALERMO,	ALFONSE
MEIDELL,	ROBERT W.	MOSIER,	ROBERT E.	PALUMBO,	ANGELO N.
MEJIA,	CARLOS	MOSNER, JR.,	JOHN H.	PALUMBO,	LOUIS
MELLIOS,	GEORGE P.	MOSS,	BRYAN	PAMPALONE,	PETER J.
MELLO, JR.,	FRANK	MOWDER,	WILLIAM F.	PANGALOS,	CHRISTOS M.
MELOIA,	JOSEPH	MOWER,	DR. ROLAND D.	PANISH,	JACK
MENELEY,	BRUCE C.	MOYER,	PHILLIP D.	PANZER,	DELMAR
MENKE,	FRANK M.	MOYER,	ROBERT A.	PAPENHAUS,	ROY
MENKE,	JACK M.	MOYLE,	WALTER G.	PARDUE,	D. EARL
MEROLLA,	GEORGE L.	MULLEN, JR.,	HARRY C.	PARKER,	CHARLES R.
MERRILL,	MILTON	MUNOZ,	PERRY	PARKER,	JOHN N.
MERRITT,	CARL	MUNOZ,	ROBERT D.	PARKER,	ROBERT E.
MERTA,	ROGER A.	MUNSON,	WARD R.	PARKER, JR.,	TERRY G.
MESSINA,	ALFRED F.	MUNTZ,	ROBERT L.	PARROTT,	ROBERT E.
METTLER,	MICHAEL	MURNEY,	ROBERT J.	PATAPAS,	EDWARD C.

PATERNI,	AL	
PATRICK,	RALPH R.	
PATRICK,	ROBERT	
PATRICKE,	JOSEPH	
PATTERSON,	WILLIAM D.	
PATTI,	WILLIAM H.	
PAUL,	WALTER L.	
PAVLIK,	DONALD A.	
PAYNE,	CWO STANLEY E.	
PAYNE,	DAVID L.	
PAYNE,	KENNETH W.	
PAYNE,	PAT	
PAYNE,	PAUL D.	
PEACE,	OSCAR	
PEARSON,	ANDY	
PEASE,	EDWARD MICHAEL	
PECKHAM,	CMDR. GEORGE E	
PEDERSEN,	VERNER K.	
PELHAM,	BRIT LEE	
PENN,	DENNIS	
PEPE,	PHILIP A.	
PERCIVAL,	DON	
PERERIA,	LIONEL C.	
PERRY,	JAMES S.	
PERRY,	SHAYLOR L.	
PESTA,	KENNETH G.	
PETERING,	ROLAND	
PETERS,	MONROE W.D.M.	
PETERSON,	GIFFORD F.	
PETERSON,	ROBERT B.	
PFEIFFER,	WILLIAM G.	
PHILLIPS,	WALTER D.	
PHILOCTETE,	EDWARD J.	
PHIPPS,	GEORGE W.	
PIERCE,	DONALD	
PIERNO,	MICHAEL C.	
PILLSBURY,	WILLIAM R.	
PINNEY,	GORDON L.	
PISANI,	LOUIS S.	
PITCHER,	WILLIAM	
PLATE,	D.C.	
PLUCINSKI,	DAVID B.	
POIRIER,	LEO	
POLINIK,	FRANK	
POLNER,	EUGENE	
POLNER,	GENE	
POND, SR.,	JAMES E.	
PONTEZ,	HAROLD J.	
POPE,	CLYDE E.	
POPE,	JACK	
POPE,	RALPH	
POPP,	WILLIAM D.	
PORTER,	CHARLES L.	
PORTER,	HAROLD B.	
PORTERFIELD,	RUSSELL E.	
POSTEN,	CHAD	
POULOS,	STANLEY	
POULSON,	RONALD T.	
POWELL,	JOHN B.	
POWERS,	JAMES A.	
PRADOS,	EDWARD J.	
PRATT,	GARDNER O.	
PRICE,	GARY L.	
PRICE,	JACK M.	
PRICE, JR.,	WILLIAM H.	
PRINCE,	CHARLES	
PROCTOR,	GEORGE T.	
PROSAK,	ALOYISUS	
PROUGH,	LEROY NELSON	
PRYOR,	FRED C.	
PRYOR, JR.,	F.J.	
PULS,	WILLIAM L.	
PYTASH,	JOHN	
QUILES,	ABEL	
QUINLAN,	PAUL A.	
QUINTO,	RONALD D.	
QUISENBERRY,	RICHARD L.	
RACE,	DALE R.	
RAGSDALE,	REAGAN T.	
RAMPY,	ROBERT E.	
RAMSEY,	RICHARD	
RANDOLPH,	WILLIAM T.	
RANGEL,	ANTONIO R.	
RASKIN,	JOSEPH D.	
RAUCH,	HARRY E.	
RAY,	FREEMAN W.	
RE,	OTTO F.	
RECKER,	LOUIS	
REDDINGTON,	MASTERCHIEF	
REDEAUX, IV,	FREDERICK	
REEDER,	DANIEL	
REICHART,	JACK	
REICHSTEIN,	WILLIAM R.	
REILLY,	MARTIN J.	
REILY,	THOMAS	
REIMERS,	FRANK	
REINHART,	JOHN JOSEPH	
REISIG,	RICHARD	
REISMAN,	CARL	
REMY MSC(SW),	KENNETH	
RENNER,	GERALD	
REYNOLDS,	JAMES F.	
RIBISL,	RICHARD	
RICHARD,	CLAYTON	
RICHARDS,	EDWARD F.	
RICHER,	ROLAND P.	
RICHER,		
RICHMOND,	MERLE	
RICHTER,	T.H.	
RICHTSTEIG,	RICHARD	
RIDDLE,	EARL J.	
RIEGO,	JOHN E.	
RIEMER,	WESTON	
RIGANO,	MRS. ANN	
RILEY,	FRED PATRICK	
RILING,	JACKSON H.	
RINER,	H. KIRBY	
RINKEL,	JAMES E.	
RITCHIE,	ROBERT L.	
RITTER,	MARLIN E.	
RIVERS,	HENRY C.	
ROBACK,	BRUCE A.	
ROBBINS,	WILLIAM PAUL	
ROBBINS SR.,	WALTER A.	
ROBERTSON,	DONALD	
ROBERTSON,	GILBERT	
ROBERTSON,	JAMES M.	
ROBERTSON,	ROBERT S.	
ROBINSON,	CHAPLAIN	
ROBINSON,	CLEMENT	
ROBINSON,	EDWARD R.	
ROBINSON,	KEVIN G.	
ROBSON,	STANLEY G.	
ROCCIA,	THOMAS A.	
ROCHA,	RICARDO J.	
ROCKETT,	CHARLES V.	
RODDEWIG,	ROBERT	
RODRIGUES,	FRANK C.	
RODRIGUES,	JOE E.	
ROESER,	THOMAS	
ROGERS,	JAMES D.	
ROGERS,	JAMES J.	
ROGERS, JR.,	GEORGE O.	
ROHM,	GEORGE H.	
ROHM,	JOHN J.	
ROJAS JR.,	MIGUEL	
ROLL,	STEVE	
ROSCHEL,	EUGENE P.	
ROSE,	ARTHUR GLENN	
ROSE,	MRS. HELEN	
ROSE,	ROBERT J.	
ROSEBOOM,	WILLIAM	
ROSS,	JOHN R.	
ROSSER,	THOMAS	
ROSSI,	PAUL	
ROSSI, JR.,	DOMINIC	
ROTH,	DAVID	
ROTONDE,	ALBERT R.	
ROTONDO,	PHILLIP J.	
ROUFF,	KENNETH	
ROUNTREE,	CHARLES L.	
ROUSE,	DONALD B.	
ROUSE,	JIM	
ROUSH,	DAVID LEE	
ROUTH,	ROBERT LEE	
ROUTSON,	RICHARD	
ROUX,	FRANCIS P.	
ROWE,	FRANCIS J.	
ROWE,	VINCENT C.	
ROWLEY,	LYLE L.	
ROYAL,	LEE R.	
ROZELLE SR.,	MARTIN L.	
RUBENSTEIN,	STANLEY	
RUBEO JR.,	JOHN	
RUBERY,	LEROY E.	
RUGGIO,	JAMES	
RUGGLES,	KENNETH H.	
RUNNEBERG,	JOHN D.	
RUPE,	SELBY E.	
RUSCITTI,	RENO	
RUSH,	WILLIAM P.	
RUTH,	DOUGLAS	
RYAN,	JOHN P.	
RYLAND,	WARREN H.	
SABO,	JOHN C.	
SABO,	WILLIAM	
SADOWSKY,	EDWARD	
SAGERER,	LA VERNE	
SAGGIONE,	PHILIP	
SAKAS,	JOSEPH	
SALARIS,	GEORGE	
SALLET,	RUSSELL	
SALTER,	THEODORE A.	
SALTUS JR.,	URBAN E.	
SALUZZI,	JOSEPH	
SALZER,	ROBERT D.	
SAMBATARO,	ROBERT	
SAMPSON,	MRS. MARGUERITE	
SANALS,	RICHARD A.	
SANDERSON,	MICHAEL	
SANDMANN,	LOUIS R.	
SANNES,	JAMES G.	
SANSO,	JOSEPH A.	
SANTO,	J. RICHARD	
SANTO,	PATRICK P.	
SANTOS,	CHRISTIAN D.P.	
SAUNDERS,	W.E.	
SAUPPEE,	J. ROBERT	
SAUVE,	ARTHUR	
SAVAGE, JR.,	VINCENT T.	
SAVELLONI,	ERMINIO	
SAVLICR,	RONALD	
SAYLES,	ANTHONY	
SCAFA,	ANTHONY	
SCALA,	MRS. JOSEPHINE	
SCALZO,	FRANK	
SCATTOLINI,	E.J.	
SCHAFFER,	WALDON C.	
SCHARF,	HAROLD	
SCHATZ,	WILLIAM	
SCHAUB,	EDWIN	
SCHEIDECKER,	ROY	
SCHEIN,	STEPHEN R.	
SCHELL,	JOSEPH C.	
SCHEUFFELE,	MARK E.	
SCHIAVELLO,	DOMEMICK V.	
SCHIESSL,	STANLEY	
SCHISSLER,	RUDOLPH W.	
SCHISSLER,	RUDY	
SCHLACK,	ROBERT W.	
SCHMIDT,	FRANK	
SCHMIDT,	LEONARD J.	
SCHMITZ,	WILLIAM E.	
SCHMUCK,	MELVIN C.	
SCHMUCKLER,	IRWIN M.	
SCHNEIDER,	NICHOLAS	
SCHOCK,	JOE	
SCHONFELD,	DAVID L.	
SCHUETTE,	GEORGE W.	
SCHULMAN,	BENJAMIN	

SCHWAB,	WILLIAM	SNAK,	RALPH	THIELE, III,	ELMO A.
SCHWARTZ,	ERNEST	SNEDEKER,	MARK J.	THOMAS,	FRANCIS X.
SCHWARTZ,	ROBERT	SNIDER,	ROBERT X.	THOMAS,	THEODORE D.
SCHWENK,	BOB	SOLBERG,	CARL T,	THOMPSON,	ANDREW J.
SCIARRETE,	ANTHONY	SOLIRA,	FELIX	THOMPSON,	ANTHONY
SCOLIERI,	JOSEPH P.	SOLOMITA,	VINCENT	THOMPSON,	J. CURTIS
SCOTT,	JON DOUGLAS	SORGE,	JOSEPH S.	THOMPSON,	JENNINGS J.
SCOTT,	MARCUS E.	SOUPENE,	LAWRENCE B.	THOMPSON,	LELAND
SCOTT,	MRS. CATHERINE	SOUTIERE,	EDMOND	THOMSON,	ROBERT W.
SCOTT,	RICHARD L.	SPOSATO,	MICHAEL	THORNBLOOM,	SCOTT A.
SCULLY	/O'DONOGHUE,ROBERT E.	SPRANGLER,	JOHN J.	THORNBURG,	CHARLES L.
SEALEY,	WILLIAM L.	SPRINGER,	CHARLES R.	THUMUDO,	MARIO
SEAVERS,	HAROLD A.	SQUAIRE,	LARRY	THURMAN,	PAUL E.
SEEBURGER,	EDWARD	ST. ONGE,	BOB	TICEHURST,	ALFRED L.
SEIDEL,	GEORGE A.	STAHL,	WILLIAM	TICKSMAN,	RICHARD A.
SEITZ,	NATHAN T.	STANGE,	DONALD M.	TICKY,	RICK
SEKAFETZ,	BRIAN C.	STANGE,	JAMES D.	TIEFENWERTH,	GENE
SELENDER,	LT	STAPLES,	ROLF	TIMM,	LOWELL R.
SELLERS,	LEWIS B.	STAPLETON,	JOHN	TISCHLER,	MARTIN G.
SELMAN,	LLOYD W.	STARCK,	THOMAS F.	TOBIN,	(USN), REAR ADM. BYRON
SELTHOFFER,	RICHARD	STARNES,	JAMES L.	TOLKEN,	FRED K.
SEMON,	GEORGE	STASSEN,	GOV HAROLD E.	TOMANEK,	ROBERT G.
SERENA,	EMMETT B.	STATON,	SCOTT	TOOMEY,	DANIEL J.
SEXTON,	PAUL	STEADMAN,	WILLARD G.	TOPOL,	CYRUS
SEXTON,	PAUL E.	STEARNS,	LARRY A.	TORCHIA,	JOSEPH
SHAFFER,	WILLIAM M.	STEELE, JR.,	JACK E.	TORRANCE,	MRS. MARY
SHALLIS,	CHARLES L.	STEFANCIN,	WILLIAM	TOWLE,	RICHARD L.
SHEA,	JOHN F.	STEIGER,	LEROY	TOWNLEY,	BRIAN
SHELLEY,	PAUL N.	STELTZ,	MILTON J.	TRACY,	NEIL
SHEPARD,	CORRIS G.	STEMPICK, JR.,	JOHN	TRAGGIS,	DEMETRIUS P.
SHIELDS,	JAMAL M.	STEPHENS,	DARREL	TRASK,	GEORGE
SHIELDS,	LAWRENCE	STEVENS,	JOSEPH W.	TRAVIS,	BEVAN E.
SHILTS,	ROBERT E.	STEVENS, III,	ERNEST	TRAVIS,	FRANK
SHIMP,	EUGENE M.	STILE,	SALVATORE J.	TRAVIS,	LEWIS F.
SHOCKCOR,	RICHARD D.	STILES,	JAMES L.	TRELLA,	GEORGE
SHOFFER,	MICHAEL	STILLWELL,	PAUL	TRIBBLE,	JAMES I.
SHORT,	MICHAEL F.	STOCKTON,	EDWARD J.	TRIGG,	KENNETH R.
SHRIVER,	LEONARD H.	STOLL,	RICHARD L.	TRIMBLE,	RAYMOND L.
SHRONTZ,	SHON	STONE,	JULIUS	TRIMM,	MRS. BEVERLY
SHULTZ,	ROBERT V.	STORACE,	TONY	TRUMAN,	JOHN C.
SHUTTLEWORTH, JR.,	VERNE C.	STORM,	LOU	TRUNDY,	CARL D.
SIBUG,	REYNALDO C.	STORY,	DAVID A.	TUCKER,	JAMES
SIEBELS,	EUGENE T.	STOUT,	BILLY J.	TULLY,	WILLIAM
SIEDLARCZYK,	ALBERT J.	STOUT,	RUSSELL F.	TURNER,	RONALD
SIEFER,	JOHN S.	STROMAN, SR.,	JACOB PAUL	TYSON,	WILLIAM
SIEGERT,	RAYMON C.	STROZ, C.S.C.,	BRO.ALEXANDER	ULCH,	DALE G.
SIELOFF,	ROBERT F.	STRUWE,	LARRY T.	UMPHLETT,	DOUGLAS A.
SIEMENS,	HERBERT	STUART,	WILLIAM M.	URBAN,	WILLIAM A.
SIEVERS,	MICHAEL	SUAREZ,	JOSE MARIA M.	URBANIAK,	MRS. HAZEL L.
SIKKENGA,	KEVIN M.	SUCHOSKI,	CHESTER	URSIDA,	DAVID
SIMMONS, JR.,	HARRY	SUFFA,	FREDERICK W.	UYEMOTO,	TOBY
SIMONS,	L.D.	SUHM,	RICHARD R.	VACCA,	DIAMANTE
SIMPSON,	ANDREW C.	SUITER,	LOUIS A.	VACCARI,	PAUL
SIMPSON,	RUSSELL B.	SULLIVAN,	JOHN P.	VAIL,	RICHARD
SKELTON,	EDWARD	SUNNE,	WALTER H.	VAN ANTWERP,	EUGENE I.
SKRODER,	SION D.	SURACI,	ANTHONY F.	VAN HAREN,	MICHAEL G.
SLACK,	DONALD F.	SURBER,	CHARLES	VAN HEE,	T.V.
SLOAN,	RUSSELL L.	SUTTON,	CARL M.	VANDERBORGH,	ROBERT M.
SLONIM,	GILVIN M.	SUVALLE,	HAROLD L.	VARGA,	MIKE
SLOTSVE,	HENRY R.	SWART,	ROBERT H.	VARNELL,	DEWAINE A.
SLUSHER,	JOHN A.	SWEENEY,	ARTHUR	VASQUEZ,	FRANK
SLYSZ,	EDWARD A.	SWEET,	KENNETH W.	VASQUEZ,	LT. JG ANDREW G.
SMITH,	A.E.	SWEITZER,	CHARLES W.	VAUGHN,	CHARLES P.
SMITH,	ANDREW J.	SYKIS,	BRUCE W.	VAUGHN,	E. BRUCE
SMITH,	CAPT JACK W.	SYLVESTER,	CAPT. JOHN	VELASQUEZ,	PHILIP
SMITH,	EDWARD G.	TAGUPA,	ALFRED T.	VELLA,	JOSEPH S.
SMITH,	FULTON B.	TAIMANA,	GIL	VENEGAS,	RALPH
SMITH,	GORDON C.	TANCREDI,	DANTE	VERDI,	BUDDY
SMITH,	HARRY	TARANTELLI,	ORLANDO P.	VETTER, SR.,	NORMAN V.
SMITH,	HARRY J.	TAUB,	GERALD L.	VICINUS,	RAYMOND G.
SMITH,	JACK E.	TAUSSIG,	CAPT. JOSEPH K.	VILLARREAL,	GUILLERMO
SMITH,	JACK EDWIN	TAYLOR,	JOHN	VILLIANO,	JOSEPH
SMITH,	JOHN C.	TAYLOR,	USN, RET., LCDR JOHN E.	VIRTUSIO,	VENANLIO V.
SMITH,	JOSEPH A.	TEDESKO,	RONALD LEE	VISCONTI,	JAMES M.
SMITH,	KENNETH L.	TEMPLE,	LEWIS DOMER	VOGEL SR.,	WILLIAM H.
SMITH,	LARRY	TERRELL,	EDWIN P.	VORHIES,	CHARLES R.
SMITH,	M.M.	THATCH, JR.,	CAPT. JAMES H.	VOSBURGH,	ALONZO
SMITH,	NORMAN	THEBERGE,	GEORGE	VULLO,	CHARLES J.
SMITH,	S.N.	THIBODEAU,	ERIC D.	WADINGTON,	EDWARD
SMITH, JR.,	CHARLES H.	THIEL,	ALLEN L.	WAGER,	WILLIAM G.

WAITE,	CHARLES G.	WHITEHOUSE,	HAROLD	WOOLSEY,	ROBERT E.
WALDRON,	EDWIN B.	WHITEMAN,	A.L.	WORKMAN,	JOHN B.
WALKER,	JAMES E.	WHITMAN,	WILLIAM	WORLEY,	WESLEY
WALKER JR.,	HENRY A.	WHITNEY,	DANIEL R.	WORM,	JAMES E.
WALKER, JR.,	WESLEY	WICH,	LAWRENCE E.	WORTHINGTON,	ROBERT J.
WALLACE,	GEORGE W.	WICKENHEISER,	PAUL R.	WRAY,	VIRGIL E.
WALSH,	JACK	WIEBER,	THEODORE R.	WRIGHT,	CAPT. GERALD C.
WALSH,	LCDR ROBERT B.	WIESER,	PAUL A.	WRIGHT,	JAMES F.
WALTERS,	ROBERT L.	WIGHT,	CONRAD	WRIGHT,	ROBERT S.
WANNAMAKER,	DON	WILBUR,	ERWIN	WRIGHT,	TIMOTHY P.
WARCHOL,	JOSEPH	WILBURN,	DENNY W.	WRIGHT,	WILLIAM
WARFEL,	HARRY	WILKIN,	VERN L.	WRIGHT,	WILLIAM J.
WARNER,	FRANK H.	WILKINSON,	KENNETH E.	WROBEL,	CHESTER S.
WASCHER,	VIRGIL B.	WILLIAMS,	ALBERT	XUAN,	NGVYEN T.
WASHINGTON,	JEROME	WILLIAMS,	BRAD G.	YAGUNIC,	JOSEPH T.
WASSERMAN,	WALTER	WILLIAMS,	DANIEL N.	YANNOTTI,	ANTHONY
WATSON,	CLAYTON W.	WILLIAMS,	H.P.	YATES,	WARREN S.
WATSON,	GUNNAR	WILLIAMS,	HOWARD F.	YEAGLEY,	CLAIR L.
WATT,	JIM	WILLIAMS,	JOHN L.	YEDLOWSKI,	MARIO E.
WATTS,	ROBERT H.	WILLIAMS,	JOSEPH	YODER,	LARRY E.
WATTS,	ZANE	WILLIAMS,	MICHAEL J.	YOST,	HENRY A.
WAUGH,	ELTON	WILLIAMS,	RODNEY	YOUNG,	HAROLD E.
WEATHERHOLTZ,	HALLER	WILLIAMS,	RUSSELL	YOUNG,	JACK E.
WEATHERTON,	ARTHUR	WILLIAMS,	STEPHEN E.	YOUNG,	RICHARD
WEAVER,	ERNEST B.	WILLIAMS,	TIMOTHY A.	YOUNG,	WILLIAM
WEBB,	BERNARD R.	WILLIAMS,	WALTER L.	YOX,	J. MALCOLM
WEBBER,	EUGENE O.	WILLIAMSON,	MICHAEL B.	YUCKA,	WALTER J.
WEBBER,	ROBERT J.	WILLY,	CHAD A.	YUDELOWITZ,	MURRAY
WECKERLY,	WILLIAM	WILSON,	LT. STEVEN	ZACCHEO,	GENE
WEIGOLD,	GEORGE A.	WILSON,	RALPH G.	ZAEGEL,	ROBERT L.
WEIMAR,	LEONARD C.	WILSON,	RICHARD A.	ZAKIAN,	JOHN
WEINER,	NATHAN	WILSON,	ROBERT	ZAMESNIK, JR.,	JOHN
WEINER,	RAYMOND H.	WILSON	EWC(SW), KIRK W.	ZECHEL,	HAROLD E.
WEINGART,	ROGER	WILSON, III,	LANSING A.	ZEIFMAN,	JEROME
WEINGART,	ROGER A.	WINGATE,	ROGER H.	ZELDES,	JACOB D.
WEIR,	THOMAS G.	WISE,	JAMES A.	ZELLNIK,	ROBERT
WEISS,	HARRY	WISE,	JULIUS	ZIENTEK,	EDWARD V.
WELDEN,	DR. R.B.	WOLF,	STEPHEN A.	ZIESEMER,	JOHN
WELLS,	ROBERT	WOLFE,	EDWARD J.	ZIKARAS,	ALVIN J.
WENGER,	JAMES B.	WOLNITZEK,	GEORGE T.	ZIMBELMAN,	ROBERT
WENYON,	LEONARD J.	WONG,	LUIS	ZIMMERMAN,	ROBERT
WESTPHAL,	MYRON	WOOD,	RAYMOND E.	ZIMMERMAN JR.,	ERNEST
WETZEL,	LEWIS M.	WOOD,	WILLIAM	ZINGALI,	BAZIL R.
WHALL,	WILLIAM C.	WOODARD,	WARRICK G.	ZIONC,	EDWARD
WHELESS,	DAVID K.	WOODCOCK,	STEVE	ZIONIC,	EDWARD
WHINERY,	J.	WOODRUFF,	WARREN F.	ZIONS,	JOHN G.
WHITE,	BERNARD A.	WOODS,	GERALD T.	ZOEBISCH,	OSCAR C.
WHITE,	JEFFREY M.	WOODS,	WILLIAM A.	ZOELLER,	CAPTAIN RAY
WHITE,	NICKY B.	WOODWARD,	FREDERICK W.	ZOLKOWSKI,	DONALD W.
WHITEHEAD,	NORMAN A.	WOODWARD, JR.,	OLIVER P.	ZUEHLKE,	KEITH ALAN

USS Missouri fires off the coast of Hawaii. Courtesy of Department of Defense/US Navy.

Association Membership Roster

Last Name	First Name	Spouse's Name
ADAMS,	ROBERT M,	MYRTLE
AHERN,	MRS. LUCY,	PATRICK
ALDRICH,	EDWARD R.	
ALTILIO,	MRS. JOSEPHINE,	NICHOLAS
AMPTHOR,	FRANCIS J.E.	
ANNETT,	MRS. YVETTE,	BENJAMIN
AUTINO,	MRS. NANCY,	GUIDO
BACHMANN,	MRS. JUNE,	JOHN J.
BAIRD,	SCOTT	
BARNES,	HAROLD T.	
BARRETT,	NATHANIEL B.,	HELEN
BAXTER,	MRS. CAROLYN,	ROBERT
BEAUREGARD,	MRS. SHIRLEY,	RICHARD
BECKER,	E.N.	
BELCH,	ROBERT D.	
BINKLEY,	WINFIELD S.	
BOHAGER,	AMBROSE	
BORRELLI,	DOMENIC J.,	ANNA
BOYLE,	MRS. TWYLA,	JAMES D.
BRANCATO,	MRS. CONNIE,	NICK
BRENAN,	N. J.	
BRENNAN,	MRS. ETHEL,	JOSEPH
BRETL,	MRS. MARY ANN,	FRED G.
BREWER,	HARRY	
BREWER,	MRS. CORNELIA,	GEORGE
BRILL, JR.,	MRS. JERLINE,	ROY
BROMLEY,	MRS. BETTY,	PETER B.
BROWN,	MRS. ELIZABETH,	RALPH
BROWN,	MRS. LUCY,	RICHARD
BUCHANAN,	C.B.	
BURKE,	JOSEPH M	
BURKHOUSE,	MRS. NORMA,	EARL
BUTLER,	ARCHIE PAUL	
CAHILL,	MRS. EVELYN,	GERALD F.
CALLAGHAN, JR.,	RADM WILLIAM M.,	BETTY
CAMARELLA,	ANTHONY	
CAMPBELL,	GEORGE J.	
CANADE,	FRANK	
CANNON,	MRS. EVELYN,	GEORGE T.
CAREY,	MRS. PATRICIA M.,	ALBERT A.
CARLQUIST,	MRS. MARY,	KENNETH G.
CARNEY,	ADMIRAL ROBERT	
CARON,	MRS. HARRIET,	DONALD
CAROZZA,	ALEXANDER,	BARBARA
CARTER,	CHARLES F.	
CASINO,	MRS. MARY,	GEORGE
CATARCIO,	SAL,	PATRICIA (D)
CELENTANO,	E. JOSEPH	
CHABOT,	MRS. SHIRLEY,	RICHARD J.
CHAPPLE,	CPL DONALD M.	
CHIBAN,	MSGT ANTHONY	
CICCOTELLI,	MRS. ANITA,	JOHN
CLANCY,	MRS. HELEN,	MIKE
CLARK,	CHESTER H.,	DOROTHY A.
COACH,	MRS. KUNIGUNDA,	BENEDICT E.
COLE,	WALTER	
COLLINS,	JAMES G.,	JOYCE
COLSON,	MRS. RUTH,	WARREN
COMI,	MRS. ROSE,	SALVATORE
CONDON,	MRS. JULIA BOYD,	WILLIAM M.
CONNIFF,	MRS. BETTY,	JOE
CONTONIO,	JAMES	
COOKE,	MATTHEW,	AGNES
COOPER,	JACOB I.	
COOPER,	JOHN	
CORDERA,	PETER P.,	PHYLLIS (D)
CORMIER,	MRS. JOAN,	NORMAND J.
COX,	IRA	
CREWS,	ELLIS P.	
CRONHOLM,	MRS. INGRID,	JOHN
DALTON,	V.	
DAVIS,	JOHN DELTON	
DAVIS,	MRS. MARY LOU,	CHARLES O.
DE BENEDETTO,	JOSEPH J.	
DE RIENZO,	MRS. JERRY,	JOHN
DEAN,	LT WILLIAM H.	
DEATON,	MRS. O.B.,	O.B.
DEGROFF,	MRS. HAZEL,	JOHN E.
DELANCEY,	WAYNE,	VELMA KUBLER
DERCOLE, JR.,	SAMUEL J.,	ELINORE
DETTERLINE,	HAROLD	
DODSON,	R. S.	
DONAHUE,	MRS. FRANCES,	FRANK J.
DONOFREE,	NORMAN N.	
DUBETSKY,	WILLIAM M.,	JOAN
DYER,	MRS. RUTH,	JAMES J.
DYESS,	MRS. DOLORES,	JAMES
EAGAN,	ROBERT J.	
ELF,	MRS. FRANCES,	ROY W.
ELGNER,	MRS. SHIRLEY,	WILBERT (BOB)
ERICKSON,	LEIF	
ERICKSON,	MRS. ALINE,	CARLTON
EVANS,	ALTON B.	
EVANS,	B.D.	
FENTON,	DAVID F.	
FEOLA,	MRS. CONSTANCE,	CHARLES
FERKAN,	MRS. LOIS,	WILLIAM C.
FIELD,	JOHN J.	
FINLEY,	WILLIAM	
FINN,	MRS. DOTTIE,	JOHN
FISH,	MRS. PAT,	JACK
FISHER,	WILLIAM J.,	MARY
FLAGG,	MRS. ANN,	HERBERT
FLEENER,	MRS. MARILYN,	BOBBY L.
FLUCK,	MRS. FRANCESKA,	THOMAS
FRANK,	SAMUEL P.	
FRENSLEY,	MRS. FRANCES,	HAROLD
FROTHINGHAM,	EVERETT	
FUSCO,	D. THOMAS	
GALLAGHER,	GARY	
GANNON,	DAVID	
GAUTHIER,	MRS. MARY,	ROGER
GERNERT,	THOMAS H.	
GIBBIE,	LESTER	
GIOVENCO,	SSGT JOSEPH L.	
GOINGS,	MRS. HARRY L.,	HARRY
GOLLY,	ROBERT	
GRANT,	MRS. NAOMI,	JOHN E.
GREENE,	MRS. DORIS,	HAROLD
GUERRIERO,	MICHAEL	
GUIDO,	JOSEPH	
GUMIENNY,	T.J.	
HADDAD,	DR. GEORGE	
HAHNE,	MRS. HELEN,	ANTHON G.
HALL,	PFC RICHARD W.	
HAMMERS,	JOHNSON	
HANCOCK,	JAMES H.	
HANLON,	JOHN P.	
HARACZAY,	MRS. WANDA,	RAYMOND
HARRISON,	MRS. FLO,	DAVID
HARRITY,	MRS. SARA,	JAMES A.
HARSHBURGER,	GLENN,	JUNE
HARTZOGE,	MRS. EVELYN,	DEWEY L.
HASSELL,	EVERETT L.	
HEALY,	MRS. FRANCES,	JAMES J.
HECK,	MRS. ISABELLE,	RAY
HECKER, JR.,	WILLIAM A.	
HENDERSON,	MRS. ELIZABETH,	CLAUDE
HEYWARD,	MRS. MARLENE,	DONALD
HILL,	MRS. PHYLLIS,	GEORGE R.
HUDSON,	MRS. ROBERTA,	HENRY
HUTSON,	MRS. PEGGY,	HUGH H.
IRWIN,	JAMES B.	
IVY,	ROY E.,	HELEN
IZZO,	LOUIS	
JACKSON,	MRS. VONDA,	JOHN A.
JAMES,	MRS. TEE,	CHARLES
JEZEWSKI,	MRS. KATHERINE	
JOERN,	HERMAN L.	
JOHNSON,	MRS. DANNA,	GEORGE
JOHNSTON,	MRS. CATHRINE,	WILLIAM M.
KANE,	ROBERT J.	

KEEGAN,	JAMES A.,	KATHLEEN
KEENE,	MRS. ROBERT A.	
KENNEALY,	MRS. MARY,	ROBERT F.
KENNEDY,	SSGT CARL	
KINDNESS,	MRS. DELLA,	RICHARD
KLAHR,	HERMAN ,	LEAH
KNIGHT,	MRS. RENEE,	ARTHUR
KOVAL,	SHERYL,	S.BAIRD(DAD)
KRZEMINSKI,	WILLIAM,	CECILIA
KRZYZANOWSKI,	MRS. ANTOINETTE,	CHESTER
KUKONU,	FRED L.	
LAFONTAINE,	MRS. MARIAN,	DONALD
LAZARUS,	MRS. INA,	LESTER
LE BLANC,	HERVE G.	
LEABO,	KENNETH W.	
LEMANSKI,	MRS. ALBINA,	ALEXANDER
LESTER,	ROBERT	
LO GALBO,	PHILIP	
LOEFFLER,	MRS. JULIA,	HAROLD J.
LOMBARDI,	CARMEN,	JOAN
LONGCHAMPS,	MRS. SHIRLEY,	JOSEPH R.
LUCID,	JOSEPH R.	
LYNCH,	MRS. GRACE J.,	JAMES F.
LYNCH,	MRS. LORRAINE,	JOHN
MACKOWIAK,	STANLEY,	ISABELLE
MALIN,	MRS. SYDNEY,	HAROLD
MANDT,	SGT JOHN	
MANN,	ARTHUR	
MAQUIRE,	MRS. GLADYS,	MAURICE J.
MARCHELOS,	MRS. IRENE,	MANUEL
MASON,	MRS. ELIZABETH,	DENNIS E.
MATATALL,	MRS. GAIL,	EDWIN
MATUSEK,	MRS. EDNA,	R.J.
MC ALLISTER,	MRS. DORIS,	J. H.
MC CARL,	R. I.	
MC CLURE,	WARREN C.,	RUTH
MC DONALD,	MICHAEL F.	
MC GIBNEY,	MRS. EUNICE,	WILLIAM
MC LELLAN,	MRS. NANCY,	WILLIAM F.
MC MENAMIN,	MRS. ELIZABETH,	NEAL
MC NALLY,	SGT JAMES	
MELTZER,	MRS. JUDITH,	GEORGE
MICAL,	MRS. ETHEL,	WILLIAM
MIHALLO,	WILLIAM ,	GRACE
MILLER,	JACK C.	
MIMS,	MRS. LORRAINE,	HOLLIE M.
MOFFE,	MRS. RUTH,	ANTHONY P.
MONKHOUSE,	MRS. MARY,	JAMES
MOODY,	MRS. MARVETTA,	DAYLE
MOORE,	MRS. BARBARA,	HARVEY
MORRELL,	MRS. LOIS	IRVING
MORROW,	DORSE	
MOSHER,	BETTY,	GEORGE
MUNYAN,	MRS. JEAN,	LEONARD
MURRAY,	JOSEPH	
NAGLE,	DONALD E.	
NELSON,	MRS. IRENE,	ALEXANDER A.
O'BRIEN,	MRS. KATE,	JAMES
O'BRIEN,	THOMAS W.	
PADGETT,	MRS. RENIE,	ROBERT R.
PAJAK,	MRS. MARY,	DOMINIC S.
PANNELL,	MRS. WYOLENE,	JOSEPH W.
PAQUETTE,	FREDERICK L.,	MURIEL(D)
PARKEY,	MRS. ESTHER,	FORREST
PEPE,	THOMAS,	JEAN
PERRY,	MRS. NANCY,	ALAN S.
PLUM,	MRS. JAN,	EDWARD
POLCHOW,	MRS. RUTH,	RICHARD
PROCOPIO,	JOSEPH T.	
RAFFERTY,	MRS. EVELYN,	BOB LEE
RAKIP,	MRS. SYLVIA,	DONALD R.
RAMSEY,	CPL THOMAS	
RAPP,	MSGT LLOYD	
REARDON,	MRS. HELEN,	JAMES F.
REILLY,	MRS. CONSTANCE,	WILLIAM
REYNAUD,	JOHN W.	
RICH,	MRS., ELIZABETH,	RAYMOND W.
RICHARDSON,	MRS. LELA M.,	JERALD E.
RIDDLE,	MRS. BENNIE,	OLAN E.
RISCH,	MRS. CHARLOTTE,	THOMAS
ROBERTS,	CPL CHARLES	
ROBERTS,	FRANK A.,	LILLIAN
ROBINSON,	CLARICE,	CHARLES D.
ROSS,	JULIUS	
ROTHERMEL,	JAMES L.,	MADELINE
ROY,	ROBERT E.,	MARGUERITE
ROY,	ROLAND,	GRACE
RUCKER,	MRS. ANNE,	JOHN
RUSSELL,	PFC WILLIAM R.	
RYAN JR.,	HENRY W.,	RITA A.(D)
RZEWNICKI,	STANLEY E.,	DORIS
SAVAGE,	RICHARD E.	
SCARPA,	MRS. ANN,	FRANK
SCHAARSCHMIDT,	HERBERT	
SCHOLL,	MRS. DORIS,	RAY E.
SCHRECKENGOST,	ROBERT M.	
SCHREIER,	JULIE,	JUNE
SCHULMEISTER,	ANTHONY,	KATHRYN
SCHULTZ,	ARNOLD F.	
SCROGGY,	MRS. SYLVIA,	LOUEL EDWARD
SEESE,	MRS. DORIS,	JOHN R.
SERVIES,	MRS. ESTHER,	LESLIE
SHAHABIAN,	MRS. FLORENCE,	ALBERT
SHAINK,	MRS. LUCILLE	
SHENK,	SSGT HENRY	
SHIPMAN,	STANLEY	
SHIPPEE,	PHILIP W.	
SIEBELS,	ED F.	
SILAS,	WILLIAM B.	
SIMPSON,	MRS. MELBA,	C. J.
SKIRATKO,	MRS. MARY,	GEORGE
SKORUPA,	MRS. BERNICE,	PAUL M.
SMITH,	ADM. HAROLD P.,	HELEN
SMITH,	MRS. MABEL,	PAUL
STINE,	MRS. ANASTASIA,	STANLEY
STONE,	MRS. JOSEPH L.,	JOSEPH
STRUBE,	MRS. DENISE,	HOWARD
SWARTZ, SR.,	MRS. LOIS,	LOIS
SWEENEY,	THOMAS F.	
SWEET,	HERBERT J.,	DOROTHEA
SWENSON,	ERICK N.,	ANNETTE
TACKETT,	MRS. SARAH,	JOHN
TARAPAIA,	MARION C.,	MARY
TATAR,	MRS. MARY,	JOHN
TAYLOR,	MRS. DONNA,	ARTHUR H.
TESTA,	MRS. FELIX,	FRAN
THOMPSON,	DUANE B.	
TOLAR,	JAMES STOVALL,	JERRI
TOWER,	EDWARD E.,	DORIS
TRAHAN,	MRS. ANITA,	RICHARD A.
TRUMAN,	BESS	
TRUMAN,	HARRY S.,	BESS
TRUMAN ,	JOHN C.	
VAIL,	MRS. SHIRLEY,	RICHARD D.
VARJABEDIAN,	VAHAN	
VOGEL,	MRS. MARIANNE,	OSWALD W.
WACLAWSKI,	CWO ZIGMUND	
WALSH,	JOHN E.	
WALSH,	MRS. TERRIE,	WILLIAM
WALSH,	THOMAS	
WARE,	MRS. PHYLLIS,	JOHN M.
WARNER,	MRS. KATHERINE,	JOHN
WARSH,	MRS. ESTELLE,	LEONARD
WATSON,	MRS. MARY SUE,	THOMAS
WESSELMAN,	DANIEL D.	
WHITE,	MRS. ANN,	CLIFFORD C.
WILLIAMS,	REYNOLD L.	
WILSON,	JACK	
WILSON,	MRS. MARIE,	RICHARD
WIRTH,	MRS. BETTY,	JOSEPH G.
WRIGHT,	ROBERT W.,	IRENE
WUTH,	MRS. DOROTHY,	HARLAN C.
YOUNG,	D. CLAUDE	
YOUST,	MRS. BETTY,	PAUL
ZARUK,	STEVE J.,	LAVETTA
ZOMBEK,	FRED	

INDEX

Biography and roster entries do not appear in the index since they are in alphabetical order in their respective sections.

– A –

Alexanderson, Elwood E. 22
Alexandia, Egypt 23
Alfaro, Ricardo 38
Algeria 14
Almond, Edward M. 33
Amoy, China 24
Anderson, Douglas 13
Annapolis 14, 17, 22, 30
Apra Harbor, Guam 12, 14
Astoria, Oregon 8
Australia 24

– B –

Balboa 10, 38
Banks 25
Bay of Argentina 35
Beardall, John R. 38
Beardall, Mrs. 38
Beaty, Quentin 23
Bird, H.V. 39
Bishop Jr., D.W. 23, 24
Blamey, Thomas 36
Bloch Sr., Donald H. 25
Boulay 25
Boylan, Michael F.X. 25
Brackin 25
Brazil 17, 33
Bremerton 8, 17, 25, 33, 37
Bremerton Naval Shipyard 8, 17
Brett, G.H. 38
Brooklyn Navy Yard 30
Burke, Arleigh 29
Burr, Donald 26
Bush, George 19
Butch, George W. "Bill" 26
Byrd, Commander 22

– C –

Cagle, Malcom W. 22
Callaghan, William M. 10, 12, 24, 27, 29, 30, 37
Carlson, Richard 30
Carmint 25
Chaho 16, 17, 40
Chernesky, John J. 18

Chesapeake Bay 10, 24, 29
Chodo Sokto 17
Chongjin 16, 17
Chonjin 22
Civil War 23
Clark, J.J. 23
Clement, Brigadier General 34
Clinton, President 17
Coccione 25
Collins, Chad H. 29
Columbia River 8
Columbia University 32
Commandant Navy Yard 10
Conners, Ralph W. 13
Cooke, Harry 29
Cosgrave, L. Moore 36
Cristobal Canal Zone 10
Crockett, Noel 13
Cuba 14, 17
Cushing, Caleb 23

– D –

Daito Shoto 12
Dalton, John H. 17
DeCarlo 25
Dell 23
Derevyanki, Kuzma Nikola evish 36
Desert Shield 8, 17
Desert Storm 8, 17, 18, 19, 33
DiBella 25
Dicks, Horace 8
Dillon, Jack 13
Donnelly, Walter 38
Doolittle, General 22, 36
Doyle, Tom 19
Duke, Captain 25

– E –

Edsall, Warner R. 17, 40
Egan, Bill 33
Eichelberger, General 36
Eisenhower, Dwight 23
Eniwetok 22
Enten 25
Europe 17

– F –

Fahr, Herbert "Herb", Jr. 4, 8, 9, 29, 30
Faulk, Roland W. 39
Fluck, Thomas F. 30
Foat 25
Ford Island 9
Foster, Orville 13
France 14, 24
Fraser, Bruce 36
Fredrikson 23
Freeman 25
Frothingham, Everett N. 13, 37

– G –

Gabney, Steve 13
Gatun Lake 38
Gatun Lock 19
Geiger, Lieutenant General 36
Getzs, Alvin 23
Gibralter 14, 23
Greece 14, 24, 25
Gregg, Donald 18
Gregory, Michael 30
Guadalcanal 22
Guantanamo Bay 14, 30
Gulf of Mexico 23
Gulf of Paria 10

– H –

Haberstroh, W.C. 13
Hagushi Anchorage 12
Haley 25
Hall, Robert 13
Halsey, Admiral William F. "Bull" 12, 14, 22, 24, 26, 29, 30, 33, 34, 35, 36, 37
Hamhung 17, 33
Hammond 25
Hampton Bays 17
Hampton Roads 14, 16, 23, 38
Hancock, Wesley 19
Hawaii 24
Hearn, Captain 38
Hearney, Richard 8
Helfrich, Admiral 36

Hill, Captain 35
Hill, Tom 8
Hillenboetter, Roscoe 10
Hitachi 13
Hodges, General 36
Hokkaido 13
Holloway, J.L. 16
Honshu 13, 14
Hungnam 16, 17, 33
Hynes, Jack 8

– I –

Inchon 33
Inchon Harbor 25
Ingram, Jonas 14
Irwin 25
Isitt, Marshal 36
Island Sea 11
Istanbul, Turkey 8
Iwo Jima 8, 10, 22, 24, 29, 30, 33, 38, 39

– J –

Jacobson, Jake 13
Japan 14, 34

– K –

Kangnung 16
Kansong 16
Kase, Shunichi 36
Katz 25
Kelley, Al 30
Kennedy 25
Kenney, General 36
Kingston, Jamica 23
Kojo 17
Kojo Wan 16
Korea 4, 16, 25, 30, 33
Korean Theater 8
Korean War 17, 25
Krueger, General 36
Kure-Kobe 13
Kuwait 17
Kyushu 11, 12, 16

– L –

Lathrop, Bob 23
LeClerc, Jacques 36
Lee, Warren S. 30
Lester Jr., Russell H. 31

Leyte 12
Libby, Ruthven 29
Limon Bay 38
Lingo, Jane Tunstall 30
Lisbon 17
Lockwood, Admiral 36
Loeffler, Harold J. 32
London, Barbara 8
Long Beach 8, 17, 19, 31
Long Beach Naval Shipyard 8, 17, 25, 33
Lourenco 25
Lovejoy, Frank 30
Loveredge, Ronald 27
Loy Sr., Maynard E. 32, 33

– M –

MacArthur, Douglas 14, 22, 29, 33, 36, 39
Majelton, Captain 38
Malone, Louis T. 34, 40
Manila 24
Mason, Frank A. 22, 25
McCain, Admiral 36
McCloskey 25
Medina, Hernando Hilario 38
Miraflores Lake 38
Missouri Mule Maud 30
Mitchell 23
Mitscher, Marc 27
Miyakazi, Shuichi 36
Mogmog Island 38
Morris 23
Morrisey, E.J. 13
Mound City, Illinois 23
Mowder, William F. 33, 40
Munson, Ward R. 22
Muntz, Bob 13
Muroran Hokkaido 13
Murphey, H.M. 13
Murray, S.S. 12, 30, 35, 38

– N –

Nadaw 25
Nagai, Yatsuji 36
Nason, Donal 25, 28
New York Harbor 8, 10
New York Naval Shipyard 8, 14, 23, 24, 29
New Zealand 24

Newell, Gordon 29
Newport News, Virginia 23
Newport, Rhode Island 29
Nimitz, Chester W. 14, 22, 34, 35, 39
Norfolk 8, 16, 25, 29, 33, 34, 36
Norfolk Naval Shipyard 16, 17
Norton, Bernice 34
Norton, Jack 33, 34
Norton, Judy 34
Nowakowski, Gene 25

– O –

O'Connor, Father 10
Okazaki, Katsuo 36
Okinawa 8, 22, 24, 29, 30, 32, 33, 37, 38, 39
Okinawa Gunto 11, 12
Okinawa Shima 10, 11, 12
Old Point Comfort 16
Omezu, General 22

– P –

Pajak 23
Panama 17
Panama Canal 10, 14, 17, 19, 29, 38
Pearl Harbor 4, 8, 10, 14, 17, 19, 22, 24, 29, 30, 32
Percival, General 22
Perry, Ensign 22
Persian Gulf 4, 8, 17, 18
Pieland, William 13
Pisani, Louis S. 34
Pitcher, William A. 34
Pius XII, Pope 10
Pohang 16
Port of Cherbourg 17
Port Said 24
Portland 31
Portsmouth, England 30
Portugal 14
Puget Sound 29
Puget Sound Naval Shipyard 25, 33
Pusan 11, 17, 18, 24
Pusan, Korea 8

– Q –

Quick, Claude 23

– R –

Radford, Arthur 27
Reagan, President 31
Red River 23
Redman, Captain 38
Reisman, Carl 35
Revels, Jack 13
Rhee, Syngman 11
Richardson 23
Riddle 25
Riemer, West 23
Riner, Hugh Kirby 13, 36
Rio de Janeiro 8, 14, 26, 33, 34
Rivers, Henry C. 39
Roberts 25
Roosevelt, Theodore 8, 17, 23

– S –

Sadowsky, Ed 13, 37
Sagami Wan 14, 35
Saipan 22
Samchok 16, 33
San Francisco 8, 17, 23, 25, 29
San Francisco Bay 10
San Pedro Bay 12
Sasebo 17
Saunders, Captain 38
Scarborough, O.D. 38
Schmitt 25
Schwenk, Bob 38
Seal Beach Naval Weapons Station 19
Seattle 17
Seoul 23
Shahan 25
Sherman, Admiral 36
Shiba, Katsuo 36
Shigemitsu, Mamoru 22, 35, 36
Shrevesport 23
Singapore 22
Smith 25
Smith, Allan E. 16, 29
Solomon Islands 22
Songjin 16, 17
Spaatz, General 36
Spruance, Admiral 22, 34
Stillwell, General 22, 36

Strait of Hormiz 8, 17
Suez Canal 24

– T –

Tacoma 31
Tanchon 16, 17, 40
Taussig, Joe 8
Thimble Shoals Lights 14
Tinian 22
Tirado, Jose Ortiz 38
Tokyo 10, 13, 14, 29, 30, 33, 34
Tokyo Bay 8, 14, 22, 24, 26, 29, 33, 34, 35
Tomioko, Tadatoski 36
Toombs, Leonard Otis 13
Towers, Admiral 36
Trinidad 17
Truman, Bess 8, 14, 26
Truman, Harry S. 8, 14, 24, 25, 26, 29, 30, 33, 34
Truman, Margaret 8, 10, 14, 24, 26, 30, 33
Turkey 14, 24, 25, 33
Turner, Admiral 36

– U –

Ulithi 10, 11, 12
Umezo, Yoshijiro 36
USS Alaska 27, 34
USS Arizona 4, 8, 33
USS Bailey 10
USS Dixie 40
USS Enterprise 11, 27, 33
USS Franklin 11, 34, 37
USS Guam 27
USS Helena 16
USS Intrepid 11, 27, 29, 33, 34, 35
USS Iowa 8, 17, 22, 24, 29, 35, 39, 40
USS Laffey 30
USS Langley 27
USS Little Rock 35
USS Macon 29
USS McNair 12
USS Missouri Association 8
USS Monitor 39

USS Nevada 8
USS New Jersey 8, 17, 24, 29
USS Nicholas 17, 35
USS Ohio 24
USS Oriskany 23
USS Randolph 36
USS Sacramento 18
USS San Diego 27
USS South Dakota 14
USS Terry 10
USS Wasp 14
USS Wedderburn 12
USS Wisconsin 8, 17, 18, 24, 27, 29, 31, 34
USS Wyoming 29
USS Yorktown 11, 27

– V –

Vallario 25
Vietnam 33
Vladivostok 22
Vorhees, Doug 37

– W –

Wainwright, Jonathan M. 22, 36, 39
Walsh, John E. 13
Webber, Robert 19
Weinberger, Casper 8
Whittiker, Mr. 13
Wilburn, Denny W. 37
Williams 25
Williams, Dwight H. 13
Wonsan 16, 17
Woods 25

– Y –

Yamato 11, 38
Yokohama 34
Yokosuka 17, 39, 40
Yokosuka Naval Station 14
Yokovama, Ichiro 36
Yongdok 16
Yung-Chang, Hsu 36

– Z –

Zellers 25

Printed in the USA
CPSIA information can be obtained
at www.ICGtesting.com
JSHW060047150824
68134JS00031B/2664